D1032471

**Simplified Guide to
Construction Management for
Architects and Engineers**

Simplified Guide
to Construction Management
for Architects
and Engineers

James E. Gorman, P.E.

President, Constructioneering Northwest Inc.
Engineers and Construction Consultants
Bellevue, Washington

Foreword by Charles B. Thomsen
President, CM Associates, Inc.

Cahners Books, Inc.
221 Columbus Avenue
Boston, Massachusetts 02116

Library of Congress Catalog Card Number: 75-34480
ISBN: 0-8436-0160-4

Printed in the United States of America

Library of Congress Cataloging in Publication Data

Gorman, James E   1940–
  Simplified guide to construction management for architects and engineers.

  Includes index.
  1. Construction industry—Management.   2. Building—Contracts and specifications.
I. Title.
TH438.G568   658'.92'4   75-34480
ISBN 0-8436-0160-4

To:
my wife, Cathy
our children, Jimmy and Colleen
Aunt Genevieve
our parents

CONTENTS

Foreword

Construction management has emerged to fill a void in the construction industry. It is a new idea; more accurately, it is a variety of new ideas conceived by many members of the construction industry to solve the problems as those members see them.

There is little argument that there is a management void in this industry, the largest in the world; however, the concepts that should be employed to fill that void are the subject of substantial argument. One of the best ways to turn a dull professional gathering into a lively conflict of opinions at AGC, ACEC, and AIA meetings is to introduce construction management into the agenda.

To some, construction management is simply a device by which a contractor may be hired for some consulting services during the design phase prior to negotiating a construction contract. Others will argue it is a totally new concept in professional services using fresh management science and technology.

The void has been filled by various traditional members of the construction industry. Contractors have developed construction management expertise, each with his own view of the services needed. Owners with continuous building programs have been forced to take heavier and heavier responsibility in managing their building projects, and in some cases, after developing the services necessary to manage their own building programs, have marketed the service elsewhere. Architects and engineers have also developed a variety of approaches to managing the project delivery process. Other companies have emerged specializing in construction management.

Unquestionably, the project delivery process requires more and more good management. Pressures on time, the increasing size of projects, the need to control costs, and our higher standards of quality require more and more attention to control techniques.

While manufacturing of building projects has created large corporations, the construction of a building is handled by an industry that is highly fragmented. An owner is faced with an enormous array of architects, engineers, and a myriad of other consultants along with contractors, subcontractors, manufacturers, and suppliers. Furthermore, few other industries have as many controls. All levels of government are involved through permits, codes, and regulations, all of which must be brought into a meaningful and related series of activities and agreements.

The construction industry is in a period of great change. Where once we had clearly defined concepts of the roles that owners, architects/engineers, and contractors played, we now have the opportunity to be as creative about the management concept as we are about the design itself. The lack of definition for construction management services is not a difficulty—it is an opportunity to be creative, to respond to the unique problems of an individual project with a unique management plan.

Jim Gorman has taken a giant step in the concept of construction management put forward on the following pages. He demonstrates a means by which all of the professional services that an owner requires may be gathered under a single contract, a concept for the construction industry which holds great promise. Anyone —contractor, architect, owner, student, manufacturer, or construction manager— who absorbs the following material will be better prepared to deal effectively with getting a building or building program completed with the tools and techniques described in this book.

Charles B. Thomsen
President
CM Associates, Inc.

List of Illustrations

Preface

Traditionally, there were three members of the construction team:
- *The owner*, who required the construction.
- *The engineer or architect*, who designed and supervised the construction.
- *The Contractor*, who performed the work.

Each was independent of the other, and each performed his necessary functions.

Today, we are seeing an ever-increasing shift of these traditional roles to the following:
- The owner.
- The architects, engineers, and construction Managers (AECM) team, or control organization.
- The prime contractor(s) or subcontractor(s).

These new roles display an ever-increasing interdependence.

Due to the complexities of modern projects, many engineers and architects already are broadening their fields and electing to provide additional services under the broad title of "Architects, Engineers, and Construction Managers" (AECM). Additionally, numerous engineers and architects are now increasing their services to include management of construction and sometimes even procurement for their clients. To accomplish this, they have increased, or are increasing, the scope of their professional services to include management and those services of managers of construction. These construction managers are new members of the architectural and engineering team and must be familiar with the basic construction processes plus recently developed management-controlled techniques. Many of the new services provided are simple expansions of the existing services, but others are new.

Ideally, construction managers have a depth and wealth of construction experience. Using such tools as Critical Path Method (CPM), Gant charts, financial reports, estimating techniques, cost controls, and practical construction experience, they endeavor to provide assistance, guidance, and supervision as necessary to facilitate the construction.

Managers of construction are expected to possess certain expertise and enthusiasm for construction not necessarily possessed by designers. A construction manager should be ready to assist in all facets of construction through the following:

- Suggesting variations in design which will result in economies and / or expedite the construction.
- Providing advice in financial and economic matters.

- Assisting in solving the problems caused by the unforeseen contingencies and exigencies of the project.
- Standing ready to provide special services.

With the advent of the management approach, consolidated under the title of "Architectural, Engineering, and Construction Management Services" (AECMS), a new team approach is being developed to facilitate the accomplishment of complex modern projects. This new AECM team approach to professional services and the organization, responsibility, and effective control of construction through the use of engineering and construction management techniques is necessary today because of the increased complexity of building projects.

Acknowledgments

The author wishes to thank his many friends and associates in the engineering and construction industry who have assisted and encouraged the preparation of this text.

Particular thanks are extended to Martha Hendrixson for her able assistance in preparing and editing this text, and to William R. Forde, Cecil W. Drinkward, C. Donald Smith, John M. Curlee, William M. Hall, and John L. Hardy for their respective parts in instructing and assisting the author.

Special thanks are also extended to the following individuals for their assistance. Larry D. Biggs, P.E., Emil J. Sederstrom, P.E., Richard C. Haas, P.E., Homer R. Castonia, P.E., William F. Serres, P.E., Gale V. Taylor, P.E., Gerald Schlatter, A.I.A./A.I.P., Theodore P. Cummings, LL.B., Franklin G. Drake, R. E. Grant, Edmund J. Gorman, LL.B., James Nelson.

Chapter 1

Introduction

The master builder is not dead. No, he has joined forces with engineers and architects, and together they are now offering their services under the title of "Architects, Engineers, and Construction Managers" (AECM). Why this alliance? Simply because it is a necessity. It is essential that everyone in the construction industry realize that teamwork gets things done, and done right. Through this alliance, the construction industry and related professions stand to gain new prestige and revenues.

The joint services of this essential trio must be adjustable to fit the needs of the owner and the project. From the very inception of a project, it is essential that professional services be supplied in this manner. A contract must be drafted that will clearly define the services to be performed and affix responsibilities. Later, if additional services are required, or the conditions of the contract change, amendments or modifications to the contract can be made.

An AECM contract must be flexible and tailored to the client's needs. The contract can include providing financial advice, including evaluations, studies, and assisting in arranging the financing. Development of a program including establishment of limits on size, facilities, and materials—also may be included in the contract.

Organization, planning, and management of the project design and construction also may be part of the AECM's job as outlined in the contract documents. In addition to reviewing, evaluating, and modifying designs to ease construction, the AECM also may provide the client a number of valuable helps during the construction process. The AECM can manage and administer the contracts with the construction contractors and subcontractors. It may schedule all phases of the construction program and furnish all architectural and field engineering services, including inspection, testing, and surveying. The AECM also can provide estimating and cost engineering estimates and prepare periodic reports for control, documentation, and cost analysis, as well as review shop and vendor drawings. And the AECM

1

contract also can include sections that require the AECM team to provide labor relations, procurement, and insurance assistance.

The contract prepared between the owner and the AECM team should clearly define the responsibilities of each party and specify what special authorities the owner delegates to the trio. The compensation to the trio for their services should be handled on an individual project basis. It should be obvious that the expanded services provided cannot be performed for the same amount of money as the design and/or supervision alone. Despite the fact that the cost of professional services has increased due to the expansion of services, the total cost of the project may be materially reduced because of the advantages which this approach offers.

Consider Figure 1.1, which represents graphically construction costs for a hypothetical project.

Note that engineering and architectural services may well cost less than what contractors are forced to allow for contingencies or finance charges. It is a rare project whose design could not be varied to provide economies and still service its aesthetic-structural requirements. As a matter of fact, there are very substantial wastes in some modern construction.

It cannot be overemphasized that, although the cost of architectural and engineering services has increased through the addition of the construction management service, the total cost of the project may well be reduced. The construction management personnel's contributions to design can lower contract prices. Their familiarity with construction techniques and problem areas often enables them to provide information which will reduce contractors' bid prices and allowances for contingencies. Through close monitoring during construction, the construction manager can also reduce costs.

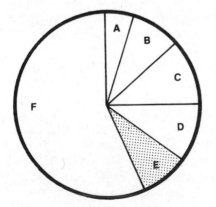

A. Finance charges by contractor.

B. Finance charges to owner.

C. Contingencies and fee of contractor.

D. AECM costs.

E. Extras and changes during construction.

F. Construction costs.

Figure 1.1 Construction costs for a hypothetical project.

Claims, extra work orders, and quantity overruns can be quite costly to owners and embarrassing to designers.

Construction managers must be unbiased professionals who are willing to render impartial decisions on a multitude of problems; they should also stand ready to assist in the total project management. Perhaps their most valuable contribution to the construction industry is their ability to make practical decisions. Although AECM services may well decrease total construction costs and wastes, the increased complexity of modern construction also demands the allegiance of these people whose training and devotion to their duties qualify them to hold positions of responsibility.

The team approach to construction was inevitable. In the past few years, technology has expanded, creating unprecedented numbers of specialists. Even the simplest construction projects benefit from their technical proficiency.

Construction managers, together with other professionals of the team, can offer expanded services, and one of the most important expansions may take place in cost controls and estimating. Accurate estimates and effective cost controls are of vital interest to AECM teams and their clients. Construction costs and estimates have been and will continue to be a prime consideration in any modern construction. Through joint effort, careful analysis, mutual understanding, and the development of an adequate program, this AECM team is able to concentrate a great deal of talent on the problems of cost controls and estimating. The development of cost estimates and cost controls is treated most extensively in Chapter 11. Let it suffice here to say that owners are reassured to know that construction management personnel will contribute to preparation of budgets, estimates, and cost controls.

Architects and engineers, by expanding their services to include managers of construction, will provide a more comprehensive service. As a minimum, this industry should benefit by having this expanded and adaptable team that is capable of acting in or managing from many positions of trust and authority. Secondly, it will benefit from cost-reducing design from the professional decisions that the construction manager's training, abilities, and experience have qualified him to make or contribute to making. Lastly, the industry will benefit from team effort in estimating and cost control. The construction manager's knowledge of construction methods and techniques and his experience will help him make important contributions to the estimating and cost control program.

The AECM team approach implies a capability, through the effective use of the talents of many personnel, to manage and design. The ways that these people can be brought together are many and the mechanics of the unifying operation can be complex, but it becomes more apparent daily that architects and engineers must resolve to work closely with construction managers.

In the AECM team approach, the construction manager is a new professional with extensive knowledge of and enthusiasm for construction. But who is he, actually? He is a member of the new breed. Members of his family include cost experts and estimators, systems analysts, computer programmers and critical path schedulers, as well as construction management specialists.

Usually construction managers acquire the special qualifications they are required to possess through education in our technically oriented institutes and/or from within the construction industry. Those who were contractors or who have been in responsible charge of construction are usually capable of making the transition from contractors to managers of construction. Construction management also draws those whose formal education has qualified them for this field. The field is also developing and training people internally, but, because it is so young, it has not been able to provide nearly the quantity of personnel that will be needed.

Construction managers possess an enthusiasm for their profession and a new and different outlook on the construction industry. They usually are practical men whose experience and education have qualified them to contribute to the design and management of construction. Their field is new, many of the techniques which they employ are new, and the demands for their services are new. They have no tradition, no established professional code of ethics, and they do not, in most cases, require license laws to assist or restrict them, though many are licensed architects and/or engineers.

Some architects and engineers have attempted to class the construction managers as semiprofessionals, but their enthusiasm and capabilities have demanded they be accepted as equal partners in the Architectural, Engineering, and Construction Management team. With proper integration of construction managers into the team, and through association and consultation with them, architects and engineers should effectively improve their professions. To work effectively, the flexible team must be organized, with responsibilities fixed and a plan of action adopted. The ability to organize this team effort into a professional organization can account for the success of many of the larger firms who offer these services.

The organization of an AECM team is, of course, dictated by the type of project and the requirements of the owner. As a minimum, it is necessary that a project team be organized under management, with subordinate groupings of professionals (see Figure 1.2.).

It is not intended that the scope of responsibilities of engineers or architects be reduced by this organization. Such organization does, however, provide an administrative head who must coordinate and direct the efforts of the members of the team.

Management could have been trained in almost any construction-related

Figure 1.2. Suggested minimum organizational requirements for an AECM project team.

profession, but it is usually advantageous to have the technically proficient and administratively capable personnel in this position. The type of project or the owner's requirements may dictate whether an architect, engineer, or construction manager will head the team. This can be shown graphically in organization charts which serve as valuable tools in forming architectural, engineering, and construction management teams. Once prepared, they show how the various people or groups involved have been organized, as can be seen in Figures 1.2, 3.1, 3.2, and 3.3.

In an organized team, the responsibilities and authorities of its members should be understood by the other members of the team. The establishment of the responsibilities and authorities of team members should be handled with care. The responsibilities of a position should be defined to include, but not be limited to, those mentioned. Authority possessed should also be defined.

The organization of any design and construction management operation should not restrict an individual's personal progress. All members of these new service organizations should be encouraged to take initiative; be enthusiastic and self-reliant; actively seek all the responsibility they can possibly handle; broaden interests and knowledge inside and outside their specialty; develop the ability to make decisions; communicate with other professionals; and contribute to the team effort.

Chapter 2

Planning

Planning the actions of the design and construction team is not quite as easy as it first appears. There are numerous ways of developing and presenting plans. Usually, in design and construction management work, the project's activities or operations and the sequence in which they are to be performed can be readily expressed graphically. Planning for design and construction services should be one of the very first steps of the project and modifications of these plans should be made throughout the duration of the project. Plans can be prepared that integrate the services of the capable team and those of contractors, or they can be developed for either one separately.

The types of plans developed and the means of expressing them can depend to a great extent on the complexity of the project. For less complex projects, plans can be graphically expressed in the form of conventional and unconventional bar charts. The more complex projects may require arrow diagrams, CPM charts, or precedence diagrams.

Usually, as a minimum, the following three plans should be developed in the early stages of the project by the AECM team:
- Project Master Services and Construction Plan.
- Project Design and Construction Management Plan.
- Project Contract and Construction Plan.

These plans are not absolutes. They serve only as guides and should be adjusted or combined to fit the needs of the program.

As mentioned earlier, planning for design and management services should be one of the very first steps of the project. This can be done by the early preparation of a "Project Master Services and Construction Plan," which can be used as a contract negotiation tool with potential clients and be included in the consultant's qualification brochures or even appended to a "Professional Services Contract." The essential function of this plan is to graphically represent a program; usually, it is convenient if this plan is not highly detailed in its original draft. In this way, service capabilities of the team are indicated and the client is allowed to investigate and evaluate the program.

A two-part sample "Master Services and Construction Plan" is shown in Figure 2.1. The bar chart portion of this plan was prepared to accommodate high rise project. Of necessity, it presents very broadly the flexible capabilities of the AECM team and can be of value in contract negotiations and general project planning.

The members of the design and construction management team and the client should be encouraged to investigate and evaluate this plan. Valuable information can be derived from it regarding the client's needs, consultant personnel, and material requirements; thus action can be initiated early to fill the needs of the program.

In the client's investigation of the plan, he should be encouraged to review carefully the consultant's flexible Master Plan and inquire into or study his own requirements and those of the project. The AECM team should be prepared to provide assistance to the owner in this investigation. Also the team should conduct its own investigations to determine its internal needs and provide for them without unduly concerning the client with any of the team's problems.

At the same time as he is investigating the consultant's planned services, the client should be encouraged to evaluate the AECM team's internal planning. In accomplishing this evaluation, the client should be encouraged to make inquiries which reveal, determine, or estimate the merit, value, effects, efficiencies, and practicality of the consultant's Master Plan. The consultant should stand ready to assist and advise the client in his evaluation and be ready to modify plans and services to accommodate the client's and the project's needs.

Through examination of the sample Master Plan, it is immediately evident that the Master Plan should be available for use during the organizational phase for all services to be provided, since more comprehensive plans will probably be prepared from the Master Plan and the service contract documents.

After the Master Plan has been adjusted and adapted to the program and a service contract negotiated, a "Design and Construction Management Plan" should usually be prepared. The "Service Contract" should define the scope of the consultant's work, fix responsiblities, provide time restrictions, and specify the method of payment. The "Consultant Services Contract" is considered later, but it should be noted here that it is the principal document in the consultant-client relationship and that it is imperative that subsequent plans be tailored to its requirements. The design and construction management team members should be conversant with the terms of the Services

ORGANIZATION | **PLANNING AND PRELIMINARY WORK** | **EVALUATION OF PROGRAM AND COMMITMENT** | **DESIGN SERVICES** | **CONSTRUCTION** | **COMPLETION**

AECM team or management define the scope of project & establish specific areas of RESPONSIBILITY.

The CONTRACT for professional services is NEGOTIATED.

The project TEAM is ORGANIZED.

PRELIMINARY INVESTIGATION AND PROGRAM EVALUATION

FEASIBILITY STUDY

FINANCIAL ADVICE AND EVALUATION

BUILDING REQUIREMENTS DEFINED

EVALUATION OF ARCHITECTURAL PROGRAM

PRELIMINARY WORK is reviewed with the client; special ADJUSTMENTS to the program or the services contract can be made at this stage. AECM team is authorized to contin. design and develop a construction management plan.

CONCEPTUAL LAYOUT PREPARED, contracts sched. prelim.budget prepared, material evaluated,cost controls estab., client reviews,approves & selects conceptual plans.

PRELIMINARY DESIGN PREPARED, site plan,sections & elevations developed, outline specs drafted, semi-detailed estimates prepared, approval by client.

FINAL DESIGN STAGE, preparation of constr. documents, working dwgs detailed specs, bid information, contact forms, scope changes eval., detail cost est, submissions to client.

Procurement initiated & vendors selected

Asst. in securing proposals & contract awards.

CONSTRUCTION monitoring, scheduling & progress control, quality & material control, monetary control including progress payments construction reports, project documentation, procedures, evaluation, cooperation with the contractors and with local authorities, as builts and job history.

COMPLETION OF BUILDING PROJ. client receives as built drawings, specs, test reports, accepts project

Fig. 2.1(a). First of a two-part sample "Master Services and Construction Plan."
(See accompanying bar chart in Fig. 2.1(b) on following page.)

8

Design/Construction Schedule: New H.1. Building/Related Facilities

PHASE	ITEM NO.	DESCRIPTION	CLIENT	AECM	SUB	1980
GENERAL	G-1	Notice to Proceed				
	G-2	Supplemental Planning		*		Compl.Summer'80
	G-3	Clearing & Grubbing		*		Compl.Summer'80
	G-4	Surveying & Mapping			*	Compl.Summer'80
	G-5	Field Investigations		*	*	
	G-6	Review by Client	*			
CONCEPT	D-1	Systems Studies		*		Compl.Summer'80
	D-2	Data Analysis & Design Criteria		*		Compl.Summer'80
	D-3	Conceptual Layouts		*		Compl.Summer'80
	D-4	Order of Magnitude Estimate		*		Compl.Summer'80
	D-5	Approval of Concept by Client	*			
PRELIMINARY		CIVIL				
	D-6	Site Work,Grading,Drainage Plans		*		
	D-7	Outside Utility Plans		*		
	D-8	Outline Specifications		*		
	D-9	Cost Estimate		*		
	D10	Design Approval				
		ARCHITECTURAL				
	D11	Building Program Scheduling		*		
	D12	Plans		*		
	D13	Outline Specifications		*		
	D14	Cost Estimate		*		
	D15	Mech.& Elec. Systems		*		
	D16	Submittal for Review		*		
DESIGN PHASE / FINAL		CIVIL & LANDSCAPING				
	D17	Site Work Grading,Drainage Plans		*		
	D18	Site Work Grading,Drainage Specs		*		
	D19	Outside Utilities - Plans		*		
	D20	Outside Utilities - Specs		*	*	
	D21	Paving,Landscap.- Plans & Specs		*	*	
	D22	Cost Estimate			*	
		ARCHITECTURAL				
	D23	Building Shell - Plans		*		
	D24	Building Shell - Specs		*		
	D25	Mech.& Elec. - Plans & Specs		*	*	
	D26	Tenant Areas - Plans & Specs			*	
	D27	Cost Estimate			*	
	D28	Spec. Systems		*	*	
	D29	Submittal for Review			*	
CONSTRUCTION		CIVIL				
	C-1	Bid Period				
	C-2	Sitework Fndtns.,Grading,Paving			*	
	C-3	Outside Utilities			*	
	C-4	Paving & Landscaping			*	
		ARCHITECTURAL				
	C-5	Bid Period				
	C-6	Building Shell			*	
	C-7	Paving & Landscaping			*	
	C-8	Building Tenant Areas		*	*	
		MECH-ELEC.& VERT.TRANSPORT SYST.				
	C-9	Mechanical Bid			*	
	C10	Electrical Bid			*	
	C11	Vertical Transport Bid			*	

Timeline columns: 1981 (J F M A M J J A S O N D), 1982 (J F M A M J J A S O N D), 1983 (J F M A M J J A S O N D)

NOTE: "ARCHITECTU-RAL" includes all civil, architectural, structural, mechanical and electrical work required within the limits of the architectural sub-contract.

C Submittal to client for civil review.

A Submittal to client for architectural review.

1 Completion of Basic Facilities.

2 Completion of all Terminal Facilities.

Fig. 2.1(b).   Second of a two-part sample "Master Services and Construction Plan."
(See accompanying planning diagram in Fig. 2.1(a) on preceding page.)

9

Contract as they apply to their fields of responsibility. The planning of design and construction services can be represented in numerous graphic ways and, in major projects, these plans can become quite complex.

Arrow diagrams are extremely adaptable to this type of planning because they can indicate interdependencies and restraints and because numerous activities can be represented in them with relative ease. This type of plan should be thorough and logical so that the logic will be available to users of the diagram and because a plan is no better than the logic it represents. The arrow diagrams may also be used for scheduling and cost estimating.

A sample arrow diagram for AECM teams is included in the Appendix at the back of this text. Projects differ; therefore the sample must be considered as only a guide. The important thing to note is that all significant activities should be included in the diagram; trivial activities should be excluded from the plan unless they are needed to further define the logic expressed in the plan.

These plans must be of concern to the leaders of the consultant team. When this type of plan or the schedule derived from it is to be reviewed by the client, it is advisable to provide an explanation and guidance for its interpretation.

In most cases, it is sufficient to explain that arrow diagrams usually read from left to right; show project activities on the diagrams; and indicate logic, interdependence, and restraints by arrows. The length of an arrow is not of importance since the arrow does not represent time. Arrow diagrams become scheduling tools if time is introduced. The schedule thus developed is only as good as the time estimates made for the activities and is dependent on the logic of the diagram. Lastly, arrow diagrams can and should allow modifications and updatings of the plan to accomodate the client's needs and those of the project.

This type of plan can be quite effective if used by all members of the AECM team, and it is usually advantageous if simpler supplementary plans are used when implementing sections of the plan. These can vary in complexity from bar graphs to mental arrangements of operations. Their intent is only to provide supplemental coordination and control to the efforts of team members.

The last type of plan to be considered in this section is the "Project Contracting and Construction Plan." This type of planning can be done in a multitude of ways. Usually, the effective AECM team's contracting and construction plans, on small or moderate size projects, can be done in bar graph form; on large or very complex projects arrow diagrams may be used.

In preparation of this plan, it is advisable to give consideration to the type contract or types of contracts to be used in the project and to the requirements of the construction project. Usually, the AECM team's purposes can

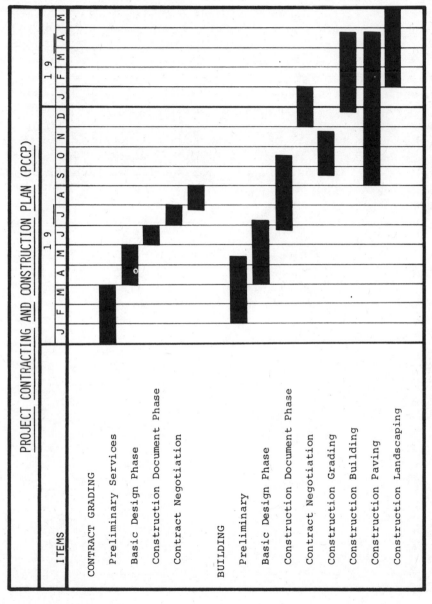

Fig. 2.2. Sample "Project Contracting and Construction Plan" (PCCP) for a small to moderate size grading and building AECM team project.

be well served by making this plan a very broad and flexible one. Improvements and expansions of it should be made during the project in a from which will satisfy the requirements of the project.

Detailed construction plans usually are prepared by the construction contractor or subcontractors. This planning, however, can be accomplished by the AECM team; but, as a minimum, they should review, make suggestions, or give opinions, especially if it is felt this will assist the contractor in doing a better job.

A sample "Contracting and Construction Plan" is presented on the following page; this would accommodate a small or moderate size grading and building project.

It should be noted that we have mentioned but a few of the plans that may be required for an AECM team project. The planning of an AECM team project is a very important operation which should be tailored to the project and in which the leaders of the consultant team must take an active interest.

Chapter 3

Organization

Organizing an AECM team consists of arranging the independent members or groups of the team into an active entity capable of performing the planned consulting function. The objective of this organization is to determine, define, and assign duties to the various members of the team so that advantages can be taken from fixing responsibilities, specialization and subdivision of work.

The organization of an AECM team will, to a great extent, be dictated by the requirements of the project. Usually, it is advantageous to have organizational planning performed on an individual project basis, so that special advantages may be taken of integration and coordination of the special capabilities of the various members or groups in the consulting team.

In organizing an AECM team project, it will become necessary to define and evaluate the functions of the various members or groups in the consulting team. Throughout the project, careful definition, analysis, and evaluation should be made and effort expended to coordinate the work of managing members with that of functional members of the consulting team.

In formulation of any AECM team, organizational effort should be spent to establish a common objective, a communication system, and a spirit of cooperation. Although the various members of the consulting team may be motivated by a multitude of reasons, their efforts must be directed to the accomplishment of the group objective. If the objective can be made more personally desirable, the members of the consulting team will more readily contribute to its attainment.

An effective communication system is a must in making the objective desirable. A free-flowing communications system, developed among the members of the team, will lead to a spirit of cooperation.

In developing a spirit of cooperation, leadership and incentive should be considered. Leadership in organizing and managing can be effectively reduced to understanding, directing, and controlling human relations in such a way as to obtain a willing obedience, confidence, respect and loyalty. In organization, consideration *must* be given to an individual's ability. The art of influencing and directing men is rarely capitalized upon.

Incentives should also be considered in organizing because men are motivated in different ways. Financial or personal incentives, or both, may have to be provided to obtain the cooperation of the specially qualified members of the consulting team. It is advisable that all members of this type of organization be afforded opportunities to: take pride in themselves and their work; relate to the organization and its objectives; and lastly, develop a sense of involvement and challenge. These aims will be accomplished by giving members of the team information, encouragement, recognition, and, to some degree, rewards.

In the formulation of an AECM team, an organization chart can be drafted to show graphically how the unit is to be established. This chart will show channels of communication from within the organization and provide information on lines of command and responsibility. The organization chart is read downward, starting from the top, or key, managerial positions. It should be remembered that the chart will represent only the formal relationships within the organization. There may exist informal relationships among the individuals of an organization which are in actual practice, sometimes stronger and more effective than formal relationships. These relationships may transcend the lines of communication shown on the organization chart.

The construction of this chart requires a knowledge of the organizational objectives. There are three common methods in use for organizing the AECM team's work. They are: organization by function, project, or a combination of function and project.

Of these three methods, organization by function is favored by very large design teams; but even in these teams, project organization and combination organization are being used more and more.

Moderate or smaller AECM teams are usually organized using the project or combination method. Examination of the samples given in Figures 3.1, 3.2, and 3.3 will assist one in making a judgment on which method of organization will best fit the needs of the project under consideration. Sufficient time and effort should be spent in preparation of the charts so that they obtain the objective of representing graphically the subdivisions of the organization and assist in achieving smooth and economical coordination and operation of the design organization.

Through the use of lines and boxes, the organizational relationships are shown in these sample organization charts. By introducing job titles, individual names, group names, functional designations, individual photographs, designations, and/or explanations of functions and responsibilities into these boxes, the organization can be observed at a glance. The lines

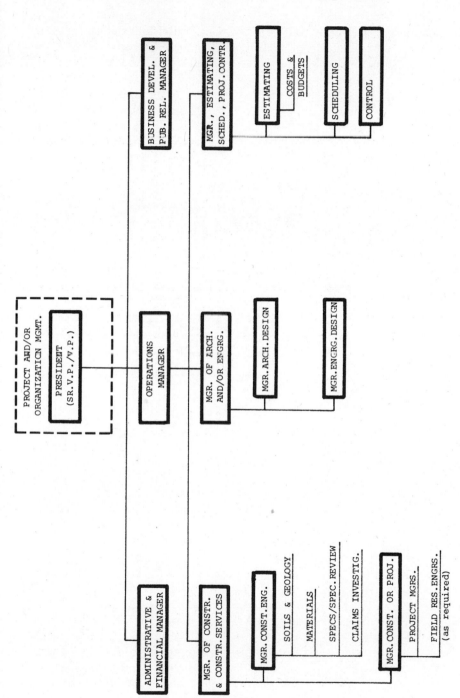

Fig. 3.1. Sample organization chart for large project --
Established AECM organization.

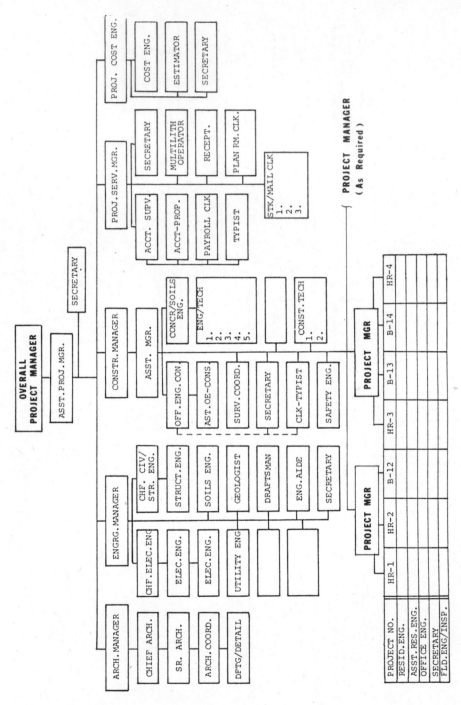

Fig. 3.2. Sample organization chart by project.

16

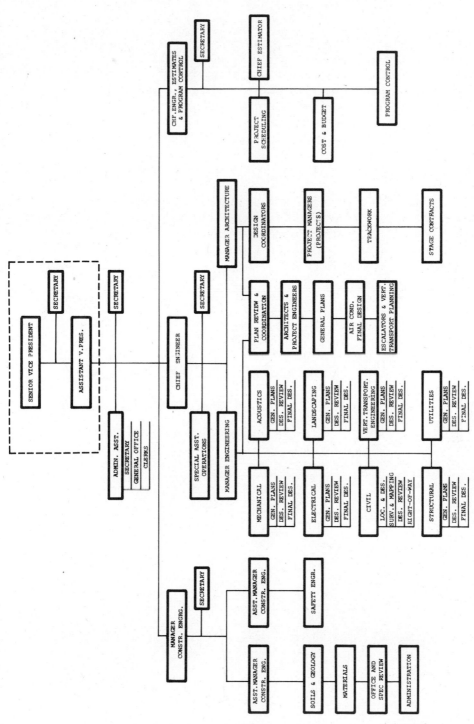

Fig. 3.3. Sample organization chart showing organization by a combination of function and project.

17

connecting the boxes represent channels of communication command, responsibility, and obligation.

The complexity of a project will have a great bearing on the AECM team's organizational adaptation for its accomplishment. Most AECM team projects should be organized using a combination of the project and function methods.

Managing individuals often have fewer than five employees reporting directly to them. Experience has indicated that this leads to better control and cooperation, though these individuals may be capable of managing far more.

A common goal leads to an effective team effort. Leadership and organizational planning will help coordinate and control the team. Only through continuing effort, cooperation, and involvement with the objectives of team members, the project, and the client, will the organized AECM team effort obtain its complete goal.

Chapter 4

Disciplines

Despite the training, devotion and, in some cases, the experience of the AECM team members, special attention must be given to interpersonal relationships. These, in turn, are integrally involved with the activity-authority relationships that exist between members of the AECM team, and between team members and those associated with the team. The relationships are partially dependent upon the differentiation of leadership, authority, responsibility, and duty.

The topics discussed in the next few pages may seem at first to be a regression into business management or human relations, but actually they are the comments and thoughts which lead to the heart of professional consulting and interpersonal relations within existing and evolving AECM organizations. Experienced professionals as well as novices would do well to give these subjects proper attention and respect.

All members of AECM teams are exposed to others and feel the effects of leadership and authority. Team members are also expected to accept responsibility and duty. The creation of an effective AECM teamwork environment is no simple task, and every individual involved must contribute to the building of the team's spirit and pride.

LEADERSHIP

Contrary to some people's beliefs, leadership is an art that can be developed; it is not an inherited trait. In almost every organized group, there will be those who aspire to control or management of the group. But these individuals are not necessarily leaders. Leaders are individuals with certain personal qualities or traits which help them gain the loyal cooperation and respect of others.

AECM work provides unique opportunities to lead others in joint accomplishment. Leaders should continuously develop and increase their ability to express personal traits, including: bearing, tack, unselfishness, loyalty, moral courage, judgment, knowledge, decisiveness, dependability, endurance, initiative, integrity, justice, and cooperation.

Perhaps the most important of these is *enthusiasm*. For AECM work, enthusiasm is essential to all aspiring leaders. Although this treatise is far too short to fully discuss leadership, one point must be made: a leader must try enthusiastically to motivate others. Although there are many motivating techniques, none will compensate for a lack of sincere enthusiasm.

Those who aspire to leadership in an AECM team are cautioned in their enthusiasm not to assume false roles or to presume that they are the only actor playing to a passive and apathetic crowd. For, in any AECM effort, very conscientious and active professionals are involved.

Aspiring leaders are also cautioned to consider others as unique, specially motivated individuals, not as people to be dominated or as adversaries. If you consider people adversaries, they inevitably have a way of becoming adversaries.

Team leaders and aspirants should be guided by and maintain the following principles: be technically competent; seek self improvement; set the example; seek authority and responsibility; develop a sense of responsibility in yourself and with your associates; employ your authority prudently; make sound and timely decisions; communicate with others and insure that tasks are understood, supervised, and accomplished; work with others; and strive for completed work.

Through this discussion, it is hoped that the reader realizes, or is reminded of, the importance of continuous effort in leadership.

AUTHORITY

Authority may be considered to have a basis in personal rights, law, and group acceptance. Through AECM work, we can have many types of authority relationships and, because of this, due care should be given to these relationships.

It is the author's opinion that far fewer conflicts would exist in AECM organizations if authority relationships were established and reviewed periodically within any AECM team. Usually, this can be effectively accomplished by limiting authority rather than by defining job functions. Every effort should be made to allow individuals to seek more authority, provided that this assumption of authority is in the interest of the team.

DUTIES

Many times in AECM work it will be found advantageous to designate actions and make specific individuals accountable or responsible for them.

Examples are numerous: the project architect may be made responsible for all building systems and layout; the cost engineer may be responsible for all estimates; or the project civil hydraulics engineer may be responsible for all drainage systems design on a rapid transit project. The advantages to be gained from such specific designations should be determined for each particular project.

RESPONSIBILITIES

Authority can be delegated, assumed, or usurped, but responsibility must be accepted. That is to say, acceptance of responsibility is a personal obligation, and one who holds an AECM organization position is presumed to have accepted his responsibility. Team members, therefore, must remember that they are accountable for every consulting act—accountable to themselves, to the profession, to their associates, and to society.

Architects and engineers have for centuries performed their respective functions and have willingly accepted the responsibilities, duties, and trusts of their professions. This should not be changed from within or as a result of AECM organizations. Through team effort, the team's functions should be subject to even more critical evaluation, review, and censorship from the many disciplines of the team members.

In AECM work, individual team members are usually quite willing to receive responsibility; many actively seek it. And, due to their unique professional and educational qualifications, they usually are quite able to accept it. Seeking and obtaining responsibility are not synonymous, however, and one of the most effective ways to frustrate a professional is to have him deluged with menial functions. It is advisable, therefore, to have the responsibilities, functions, and obligations of each member of an AECM team handled on an individual project basis, with consideration given to the individuals and the project. How and to whom responsibility and its accompanying authority is delegated has a marked effect on the organization's efficiency. If this delegation fails, those who have delegated must recognize the failure and apply corrective measures.

The ability to delegate authority is not possessed by all men. Effort should be made by all individuals who delegate responsibilities to develop character traits which will allow them to more easily obtain confidence, respect, and cooperation from subordinates. The traits to be developed include enthusiasm, bearing, integrity, tact, and judgment. Even if the person delegating the authority possesses all of these, and other traits besides, team members should not be expected to perform functions or accept responsibility for work which has not been clearly communicated to them. Architects, engineers, and construction managers are almost invariably able to start themselves and produce what they "think" is desired. It is imperative that lines of communication be established and clear instructions given, so that duplication, error, poor coordination, and ineffective use of professional talent be avoided.

AECM organization members should be stimulated by their work functions and responsibilities. They should be encouraged to cooperate in both individual and group effort. From this encouragement two benefits may be realized. First, the creative ingenuity and inventiveness of the team members will be directed toward common objectives. Secondly, by becoming intimately involved in the effort, the individual associates with the team and acts as a "consultant" or "partner," rather than as an individual. This assumption of the role of consultant or partner will psychologically motivate and reinforce the individual, so that he can enthusiastically support the team effort.

Responsibility brings with it obligation and decisions. Obligations must first be recognized, then serviced. Usually, organizational obligations can be kept separate from individual obligations, and it is usually advantageous to keep them separate. If this cannot be done, a decision to serve either organizational obligation or personal obligation may be forced on an individual.

This type of decision is very difficult to make, so we are led naturally to the old but invaluable techniques of: defining the problem; determining the objectives; reviewing all of the facts and weighing the alternates, including testing possible action, if practical; making the decision and then taking action; and checking results.

Through a logical process such as the one indicated above, which is almost a necessity in professional action, an individual is able to organize his thought and decision processes and is thus aided in making his decision. If it later becomes apparent that he has erred, he must also be man enough to take corrective action; his ability to do so is one of the greater proofs that he has indeed accepted responsibility for his actions.

Obligations and decisions imply authority, and, in well planned and organized AECM team effort, authority must be possessed and delegated. Each member of the team must have the authority to accomplish that for which he is responsible. The individuals forming the consulting team are almost always uniquely qualified to perform their functions and to assist in the accomplishment of the team's objective. Because of this, they are extremely sensitive to authority and authoritarian type action.

There is no more effective way to destroy an AECM team effort than to force a one-man organization or authoritarian control upon members of the team. This is not to say that strong individual leadership is not adaptable to this type of group effort; in fact, it is extremely effective, especially for small and moderate size project organizations.

It should be the objective of all managers or leaders in this type of organization to have fellow team members feel that they share in the responsibility for execution of the program.

Human relations and authority are uniquely linked; therefore, in exercising authority in this type of organization, one must never forget to treat people as individuals. Authority should be used as a tool to create better human relationships. In AECM teamwork results can only be obtained through people. A person with authority in the AECM team must encourage others to work for and with him. To do this, one should figure out precisely what he expects from others and clearly and concisely communicate it to them.

He should also point out ways of improving the performance of his fellow workers and his subordinates and of letting them know how they are progressing. Those in authority should be willing to give and share credit when it is due. Compliments for extra or exceptional performance will usually bring out the best from AECM organization members.

Secretiveness is not a mark of authority. Every effort should be made to keep the team informed in advance of changes, along with the reasons for the changes, if possible.

A conscientious effort must also be made by those with authority, or those who aspire to authority, in AECM team work to find ability not in use and to make use of it, or at least to make plans for using this ability. In this light, those possessors or potential possessors of authority in an AECM team are cautioned not to stand in a man's way. Opportunities, advancement, and change are essential to many men, and resisting this will only divert their effort from the team's objectives or frustrate them.

Responsibility, duty, and authority are very closely related in AECM work. Responsibility indicates an obligation and accountability for the accomplishment of a function, task, or objective. In accomplishing these functions, tasks, or objectives, a team member incurs duties and obligations to organizations, individual persons, and self. To accomplish an objective, authority will have to be possessed and properly exercised. Authoritarian action is intolerable in AECM team work, where results are so dependent on individual accomplishments.

Chapter 5

Selection of a Consultant

The type of project should dictate the type of consultant selected; therefore, we will discuss in this chapter certain variations of the basic AECM team organization. These variations and adaptations permit the basic team to provide additional management services to the client, as required for the particular type of project under consideration.

Various nomenclatures, abbreviations, and designations for the essential AECM service can be set up to assist in specifying and understanding what service is being rendered by the AECM team or firm. Although unlimited possibilities exist in setting up such designations, let us direct our attention to the following three examples:

MAECM = Management for Architectural, Engineering, and Construction Management.

AECMM = Architectural, Engineering, Construction Management and (Additional) Management (as required).

AECMP = Architectural, Engineering, Construction Management and Procurement.

Obviously, numerous other variations can be developed, and these abbreviations are often helpful in specifying what the AECM team is to do or how it will operate for a client.

If we consider a potential project situation, we may be able to see how these designations could be applied advantageously by an owner, agency, or potential consultant. It is also desirable for the owner or agency to designate what services are required from the consultant and for the consultant to designate the type of services he is able to render.

To better understand this and the whole AECM consultant selection process, consider the following hypothetical case. Center City University is a medium sized privately owned institution located on the West Coast, with an enrollment of nearly 5,000 students. The university is situated on what appears to be a fully developed campus and has just been given a new

80-acre campus site located ten miles away from its existing campus site.

Traditionally, individual authorities, advisors, and consultants, working independently or under administrative or other direction, would establish a program for the development of the new site. Then, through multiple contracts with architectural firms, engineering firms, and others, the university would provide for the development.

Because of the nature of C. C. University's problems, it seems apparent from the very beginning that an AECM team management approach might prove highly effective, since the team approach provides organizational unity in planning, organizing, designing, controlling, and directing this development. Notably, MAECM consulting services could be suggested for or sought by the university. The use of an AECMM team also seems natural and highly advantageous to the university. Once created, this professional consulting team could: work with the directors of the university and any other involved groups to perform the preliminary work of preparing an evaluated plan, program, and budget; prepare conceptual layouts, budgets, schedules, and programs; perform integrated architectural and engineering design programs; prepare and administer construction contracts for the university; and provide assistance in operating, training, and administering in the completion phases of the university development.

Assuming that C. C. University is administered by a university president who works in conjunction with a board of regents, how would they approach the multiple problems of this development? Some of the preliminary work they should do is: define in general terms their specific situation and their objectives; determine in general what will be the limiting factors, such as time, money, political opinion, etc.; seek professional advice, preferably through retention of a MAECM (or AECMM) consultant.

The mechanisms used in selecting the MAECM (or AECMM) consultant are essentially the same as those for selecting consulting engineers or architects. The initial procedure usually is a written invitation to AECM firms to submit qualification brochures which indicate their capabilities to perform the requested services.

Assuming that Central City University would prefer to work with only one consultant—which, in the author's opinion, is the best way to handle this type of development—the invitation to submit a qualification brochure should specify in as much detail as possible what services would be required. Generally, however, it will suffice to indicate that the MAECM (or AECMM) consultant team is to provide the following services:
- Total project management or parts thereof.
- Master planning and feasibility studies, including budgeting or conceptual estimating.
- Preliminary design.

- Detail design.
- Management of construction.

An expression of the extent of service that the consultant could render should definitely be sought, particularly if the quantity of work or services could be expected to increase during the project.

Upon receipt of qualification brochures from those consulting firms which elect to submit their qualifications, the University's administrators should critically evaluate these qualifications. An oral presentation of the consulting firms' qualifications might also be made in conjunction with the presentation of the written brochures. Usually it is best, if more than four firms are being considered, to first receive their written qualification brochures for evaluation. Then, on a selective basis, certain of the firms should be requested to make an oral presentation of their qualifications. Customarily, three firms are requested to make oral presentations on a project.

The selection reviews and evaluations should be conducted as rapidly as possible, and the consultants should be advised of their status.

After the evaluation discussions, a service contract may be entered into with the selected consultant firm. Discussions regarding the service contract particulars, including fees, are appropriately considered at this time.

Some owners and agencies find it helpful to conduct selection procedures in the following manner. First they go through a general selection process. After they've picked the ones they like best, the owners send invitations to a group of qualified consultants to submit qualification brochures. Oral presentations also may be requested, though the first oral presentation is not essential. Brochures and presentations by all consultants should be received during a specified time period.

With the qualification presentations in front of them, the owners evaluate each consultant's qualifications and determine the three who seem best qualified.

After advising consultants of their status, the owners usually ask for second presentations by the three preselected consultants. This step may not be used, but it does give those selecting the consultant a second chance to evaluate all three.

Then consultants are ranked as first, second, and third. (This information *may* be given to the competing consultants.) Finally, the owner sits down with the first choice to discuss contract terms. Fees and other contract particulars should be discussed at this time. If satisfactory terms cannot be reached with the first-choice consultant, the second-choice consultant may be invited to discuss contract terms, and so on until the owner finds what he's looking for.

Adherence to such a schedule is easy, since these procedures are relatively simple, but selection of a consultant is not an easy operation. Due consideration should be given to the reputation, capabilities, and experience of the consulting firm. The qualification brochure and oral presentation should provide much of this essential information about the firm. Certain owners or agencies request specific proprietary, policy, and financial information from prospective consultants.

Although qualification brochures may vary in complexity and style, there are certain elements which should be included. Figure 5.1 provides a sample list of the contents of a typical written presentation of qualifications. This sample could also be used as a guide for those preparing proposals, so that they would include the more pertinent information in their qualification brochures.

The sample list in Figure 5.1 is not intended to be all-encompassing, but it does provide an easy reference for information that should be obtained and evaluated from most qualification brochures.

The next evaluation is that of the oral presentation of the consultants. This evaluation is possibly the hardest, because the most capable consultants may not be the most eloquent, and the evaluators may not have been technically educated. In general, however, this presentation should be viewed with a critical eye. Remember that a qualified, ethical, and capable consultant is being sought.

The author also wants to caution those who are to evaluate a consultant's presentations not to assume that age is a mark of capability. A good evaluator will try to determine what education and experience the prospective consultants and their organizations have had and really evaluate this in relation to the project at hand.

Some things to look for during a presentation include professional attitudes, mental alertness, enthusiasm, and bearing. The author has found it advantageous to make brief notations during many presentations to aid in the recolleciton of the presented material. The use of a ''Presentation Evaluation Sheet'' has also been found helpful in rehearsing for oral presentations or constructively criticizing colleagues making presentations. A sample presentation evaluation sheet is provided in Figure 5.2. This sample is not suggested for use by those receiving oral presentations of professional qualifications; however, it can be used to the advantage of AECM team members in evaluating or rehearsing their own oral presentations.

These evaluation criteria are particularly addressed to speaking and speech techniques, a place where so many competent AECM team members can fall down. It is presented here as an aid to those who will be making this type of presentation.

The value that is placed on oral presentations varies quite a bit, but a good presentation is a mark of the true professional. It is reasonable to assume that a presentation may be the determining factor in selection among qualified consultants.

Professionalism is the key word for both the potential client and the

QUALIFICATION BROCHURE
FOR A MEDIUM COMPLEXITY
MAECM (OR AECMM) PROJECT

Contents   **Page**

Letter of Transmittal (not necessarily in brochure) i
Table of Contents and List of Illustrations ii
Introduction and Summary of Contents iii

Section

I.   Organizational and Proprietary Information

II.   Subcontractors

III.   Ability of Consultant to Meet Requirements of the Program

IV.   General Information
          A.   Name, Address, Telephone
          B.   Type of Firm
          C.   Organization of Consulting Firm
          D.   Financial Information

V.   Statement of Consulting Experience

VI.   Personnel and Organization Information
          A.   Organization Charts
          B.   Key Personnel
          C.   Resumés

VII.   Statement of Administrative Policies and Practices
          A.   Personnel Policies
          B.   Financial Policies
          C.   Project Control and Scheduling Policies

VIII.   Statement of Interest in Project

IX.   Additional Information
          A.   Cost Improvement Program
          B.   Quality Assurance

Figure 5.1. Sample of the contents of Qualification Brochure for a
MAECM (or AECMM) project of medium complexity.

PRESENTATION EVALUATION SHEET

BEGINNING: DOES THE SPEAKER . . .

Begin with confidence and assurance?	Yes	No
Get your attention immediately?	Yes	No
Establish purpose and provide subject information early?	Yes	No
Articulate clearly and precisely?	Yes	No
Speak loudly enough?	Yes	No
Speak too loudly?	Yes	No

MANNER AND DELIVERY: DOES THE SPEAKER . . .

Move awkwardly while speaking?	Yes	No
Display distracting mannerisms or habits?	Yes	No
Fidget too much?	Yes	No
Seem genuinely interested in his subject?	Yes	No
Seem genuinely interested in his *audience?*	Yes	No
Handle visual aids well? If not, give reason below: (Speaker talked to visual aids; presented them too rapidly; poor visibility; too technical; etc.)	Yes	No
Use grammar, pronunciation, and vocabulary in a style suited to the audience? If not, give reason: (Style is too formal; too technical; frivolous; colloquial; etc.)	Yes	No
Speak too rapidly?	Yes	No
Speak too slowly?	Yes	No
Speak at an irregular pace?	Yes	No

MAIN PART OF PRESENTATION: DOES THE SPEAKER . . .

Hold your interest?	Yes	No
Proceed in logical sequence?	Yes	No
Stick to subject of presentation?	Yes	No
Indicate a lack of preparation in certain areas? (Indicate which areas.)	Yes	No

ENDING: DOES THE SPEAKER . . .

Summarize well?	Yes	No
Leave you confused on anything? If so, what?	Yes	No
Invite queries or requests for clarification?	Yes	No
Conclude with finality?	Yes	No

Figure 5.2. Sample presentation evaluation sheet for use of AECM team members in rehearsing oral presentations.

consultant. Competition on the basis of changes, or the maligning of another's reputation, either directly or indirectly, is not only unethical but is unsatisfactory conduct which should not be tolerated. Rather, through an adapted adherence to the procedures set forth in this chapter, selection of an AECM consultant should be facilitated.

Chapter 6

AECM Services & Contract Components

It is essential that Architectural, Engineering, and Construction Management services be kept flexible, since they must be adaptable to almost all types of construction projects and almost all types of clients. Generally, through a service contract, the consulting team and the client agree as to what services will be provided by the AECM team. An agreement may be entered into which designates the AECM team as the "owner's agent" or "representative." Thus, the team can be made responsible for performing all architectural and design functions; all engineering; and all the arranging, scheduling, coordinating, and supervising of the project. The client can participate in these functions to whatever degree is desired and agreed upon.

By empowering the team to perform, the client can designate the AECM team as the one entity with authority to assume responsibility for *all* of the following functions: Planning; architectural, engineering and, construction administrative functions; scheduling and estimating; equipment and materials purchasing and inspections including expediting all deliveries; design, including the preparation of all specifications, drawings, and contracts; all construction and supervision though it is usually advisable for the AECM team to supervise the construction contracts by having the work performed by independently managed construction contracting firms; and preparation, negotiation, and execution of changes.

In almost all cases, economies will be realized by having the team perform all of these functions, while the client acts as monitor and controlling agent. Delegation of such authority to the single AECM team will reduce duplications and confusion.

A few words of caution are necessary here. First, to the owner: Approach the problem of selecting an AECM team with the same spirit as you would in selecting any other professional group. Choose the professional help you need and then be prepared to stand with the team through to project

31

completion. The professional reputation of the team and team members should be a strong influence in the selection of an AECM team, since it may be assumed that the team and its individual members will strive to preserve and enhance their reputations.

Secondly, to the AECM team: Remember that if your team is retained, you must retain its identity as an impartial and unbiased professional unit. Team members must also stand ready to meet and overcome all problems caused by their own errors in judgment or forecasting. There should be no attempt made to sidestep, deny, or disclaim responsibility for errors of judgment or for unforeseen circumstances. The owner should be warned that he will stand the financial costs resulting from errors in judgment by his consultant or for unforeseen circumstances, provided that the team has not been negligent.

The team must strive to be an unbiased professional arbitrator. Relations may tend to become strained between the owner and the consultant, or between the consultant and contractor(s) or subcontractor(s), if errors of omission are found in the plans and specifications. In the event of difficulties, all interested parties should try to remember that few human accomplishments are perfect. And, of course, the consultant should not attempt to require the contractor to bear all risks without regard to circumstance.

Because of the many different types of projects and the various needs of projects and clients, an individual service contract should be drafted for every AECM project. The key word for this contract preparation should be "adaptability." Usually, the consultant will prepare the contract; the use of semistandard format is not uncommon, to which may be appended special section definitions, and descriptions.

The preparation of an AECM service contract can be quite complex and is generally more detailed than what this treatise can give. It is advisable to obtain the assistance of an attorney whenever necessary in preparing AECM contracts. Make sure that the words of the contract express clearly, concisely, and explicitly the intent of all parties to the contract.

Adopt a format for the contract which is appropriate to the project. Often, this type of contract can be divided advantageously into these three parts: Contract Agreement section; General Terms and Conditions section; and Special Conditions section.

Engineering and architectural associations will often provide sample contracts for services which, with certain revisions, can be made to accommodate AECM projects. Legal advice and assistance is advisable for preparing AECM service contracts on moderate or large size projects. A complete sample AECM contract is presented in Chapter Seven for reference.

CONTRACT AGREEMENT

The Contract Agreement section of an AECM service contract may be organized to include the following:

A. Contracting Party section.
B. General Terms and Conditions Reference section. General terms and conditions are made a part of the agreement in this section, and a copy of these general terms and conditions may be appended to the agreement.
C. Project Description section.
D. Scope of Work section, which explains with or without supplements or appendices, exactly what "services" are included in the contract.
E. Costs section, including a description of what shall be done to control design and construction costs.
F. Schedules section, with dates established (by the AECM team or the client) for for design reviews, design completion, and construction.
G. Compensation section, which will establish how the consultant is to be reimbursed for his services.
H. Entire Agreement section, which defines what documents are part of the agreement and specifies what services performed or caused to be performed by the AECM team prior to the date of execution of the contract shall be deemed to have been performed under the contract.
I. Signature and Witness section.

It is usually convenient to have the Contract Agreement section precede all other sections of an AECM service contract, because it contains the basic components of the contract. The other sections of the contract will usually complement, elaborate, or expound upon this section. Thus, the General Terms and Conditions section and the Special Conditions section, along with all supplemental schedules and exhibits necessary, can conveniently follow the Contract Agreement section.

GENERAL TERMS AND CONDITIONS

The General Terms and Conditions section contains what some consultants refer to as the "boiler plate," probably resulting from the repetitive nature of the information contained within it. General Terms and Conditions generally can be expected to include sections on the following: Definitions and General Terms Explanations; Client and Consultant Status; Control and Progress of Work; Payments, with any schedules that complement this section; Responsibilities and Limitations; Audit or Consultant Costs Verification; Insurance; and Force Majeure and Notice.

SPECIAL CONDITIONS

This section deals more specifically with the project. The scope of the consultant's services should established clearly and explained in detail in this section. Usually, the unique conditions and special services required by the client or project are enumerated here. Special Conditions can also include the following subsections under a title, "Scope of Consultant Services": general explanation; design and plan preparation descriptions; a sequence of services program, including phase definition and/or schedules; a plan preparation program; special design or management services explanation and designation, and statements indicating the phases in which these programs are to be provided; work schedule; section on work by others; and section on cooperation with others. Supplemental schedules and exhibits may be appended to these sections.

After drafting, review, negotiation, and execution, the AECM service contract becomes a primary tool in the client-consultant relationship. It is extremely important that the client and his personnel, consultant team managers, and team members be familiar with the contract as it pertains to their jobs.

AECM service contracts usually will cover many highly technical services; therefore, the contract and negotiations should lead to the creation of an environment of mutual trust and confidence between the client and consultant. Creation of such a working environment will be facilitated by the preparation of a sound legal document and through mutual understanding. Someone thoroughly conversant with AECM services should supervise the preparation of the contract—particularly of those sections describing services to be performed, sequence of work, information to be furnished by the client, and terms of payment.

AECM teams, engaged on the basis of their qualifications and experience, are compensated on the basis of negotiations. AECM services are not suitable for competitive bidding, because neither quality nor quantity of professional services can be neatly defined. A client who attempts to buy AECM services by competitive bidding will have no realistic basis on which to compare the value of services. The employment of an AECM team on a "contingency basis" is also considered unsatisfactory. This contingency basis refers to a practice of requesting AECM firms to submit preliminary reports and estimates without charge in the expectation of being retained as AECM consultants *if* the project is undertaken. The dangers of this practice to both client and consultant are quite evident, because a favorable recommendation for the project under investigation is made with, or as a condition of, compensation.

Charges for AECM services can be based on any one or any combination of the following methods with appropriate modifications applicable to the specific case: per diem, retainer, and salary costs times a multiplier plus direct nonsalaried expenses.

Per Diem

Charges for the personal services of individual AECM team members can sometimes be based on a daily rate. In general, this rate will vary from $200 to $1,000 and upward per day for principal consultants. In addition, reimbursement for travel, subsistence and other out-of-pocket expenses may be required. This method of compensation is particularly useful in preliminary study, evaluation, and legal work and for direct personal services supplied on an intermittent basis.

Retainer

The employment of an AECM team on a retainer basis offers some unique advantages. A client may retain the multidisciplined consultant team or he could retain individual team members in this way.

The amount of the retainer and the terms of agreement for services on a retainer basis vary widely. Compensation may be based on a fixed sum, paid monthly, or paid on some mutually agreed upon basis. Through the retainer, the client is assured that the professional advice and service will be available to him on call.

Salary Cost Times a Multiplier
Plus Direct Nonsalaried Expenses

This method of determining charges is the most favored because of the complexity and nature of AECM work. For most AECM projects, it is impossible to state accurately the scope of the work at the time the team is negotiating a service contract. Preliminary Architectural, Engineering, and Construction Management work is usually so indefinite at this time that no other basis of charges would be satisfactory or equitable.

In this method, salary costs are defined and a multiplier applied; and direct nonsalaried expenses are defined and a service multiplier applied. The sum of products of these multiplications is then charged to the client.

Salary Cost is usually defined as the cost of salaries (including sick leave, vacation, and holiday pay) applicable to specific AECM team members for the time they spend that is directly chargeable to the project, plus unemployment and social security contributions, excise payroll taxes, employment compensation insurance, retirement benefits, and medical insurance benefits. Salaries or imputed salaries of principals or partners in an AECM team are usually considered chargeable salary costs to the extent that the principals or partners performed technical or advisory services directly applicable to the project.

The Multiplier, which is applied to salary cost, may vary from 1.5 to 3.0 times the salary figure. This factor is to compensate the AECM team for overhead, plus a margin for contingencies, interest of invested capital, readiness to serve, and profit. Larger multipliers are generally required on smaller projects, although the multiplier may vary upward with the experience and capabilities of the AECM consultants.

Direct Nonsalaried Expenses are those expenses incurred by the consultant team as a result of a project's operations, exclusive of overhead. They generally are considered to include services directly applicable to the work—such as special legal and accounting expenses, computer and programming charges, laboratory expenses, printing and binding expenses, and similar costs that are not applicable to the general overhead. Living and travel expenses of employees and, sometimes, principals or partners, are also a part of direct nonsalaried expenses. Identifiable communication expenses, as well as drafting and stenographic supplies and expenses, and reproduction costs chargeable to the client's work are also covered. These expenses, which seldom can be determined in advance, are charged to the client at actual invoice cost plus a nominal service charge of 10 to 25 percent.

Generally, AECM charges will be based on one of the three methods of compensation outlined above. The last of the three, Salary Cost times a Multiplier plus Direct Nonsalaried Expenses, probably provides the most equitable arrangement for this type of work. Once a client and consultant team have agreed on the team's basis for charging, controls may be placed on these charges through a Compensation Agreement. Following such agreement, these charges are often referred to as "costs." Many agreement forms may be adopted, but compensation is usually controlled through contract provisions which specify that the consultant firm is to be compensated for AECM services by one of the following methods: Costs plus a Fixed Payment; Fixed Lump Sum Payment; or Costs times a Multiplier, limited by a percentage of the construction and total project costs as well as cost estimates. These charges for AECM services are usually made within the framework of one of the following types of contracts: Cost plus a Fixed Payment Contract, Fixed Lump Sum Contract, Cost plus a Fee Contract, or a Percentage of Costs Contract.

Costs Plus a Fixed Payment

Compensation made by the Cost plus a Fixed Payment method offers a satisfactory basis for the performance of AECM services. The contract should contain the explicit statement of what costs are. Notably, in this use, costs often refer to: per diem plus expenses; salary costs times a multiplier plus direct nonsalaried expenses; or salary costs plus specified overhead costs plus direct nonsalaried expenses.

It is advisable that the client know exactly what charges are considered costs in this application.

Fixed Lump Sum Payment

Compensation for AECM services by a Fixed Lump Sum Payment is satisfactory only when the scope of the assignment to be undertaken can be clearly and fully defined. This type of agreement is not favored by most AECM consultant team members even when such a full definition can be made. These specialists rightfully consider Architects, Engineers, and Construction Managers highly professional and almost always believe that AECM teams are worthy of hire on a service basis that does not lead to economic comparison of professional capabilities. Clients are cautioned that quality and quantity of technical services cannot be neatly defined, and, through restriction of AECM expenditures, they may materially increase the total project cost. They should also be reminded that other cost control mechanisms are available.

Percentage of Costs Payment

Compensation by Percentage of Costs may be based on costs and limited by a percentage of construction costs, construction cost estimates, total project cost, and the total project estimate. This method of compensation is widely used for both architectural and engineering work, and its modification to accommodate AECM work should present little problem. This method assumes that there is a correlation between the AECM service costs and construction costs that may be calculated satisfactorily in a simple linear equation.

It is true that AECM service costs can be limited in this way, but the reader is warned not to assume that the AECM talent required to perform services will remain constant for similar projects over a period of time. The requirements for manhours of technical talent in design and construction have risen consistently in the past few years. Automation, computers, and improved techniques have, and will continue to have, an effect on AECM talent requirements. Further, it should not be assumed that the cost of AECM services will increase at the same rate as costs of construction.

Obviously, the problem reverts to determining what percentage of construction costs, construction cost estimates, total project costs, or total project estimates will limit the AECM team's fee. This percentage will depend on many factors, including project size and complexity, time restrictions, team reputation, and many more. In general, it should not be uncommon for negotiated AECM charges to run from 7 to 15 percent of the bases mentioned.

Usually, the agreement reached between the client and the consulting team can be appended to the service contract. It may be advantageous to append two schedules, one which clearly explains what are chargeable costs, and the other referring to payments. Other schedules and exhibits may also be appended to AECM service contracts. Progress schedules, insurance policies, and procedure requirements can often be handled in this way.

The preparation and negotiation of AECM service contracts can be extreme. Legal advice and technical assistance can generally be sought by either party, but an environment of mutual trust should prevail. AECM service contracts may consist of three parts—the Agreement section, the General Terms and Conditions section, and the Special Conditions section—and may include appendices containing additional exhibits and schedules. An important part of AECM service contract negotiations involves payments and charges for services. There exist various ways of making compensation to a consultant team. The important thing is that an equitable agreement must be arrived at and made a part of the service contract.

Lastly, it is imperative that all parties to an Architectural, Engineering, and Construction Management service contract be thoroughly familiar with the details of this primary document.

Chapter 7

The MAECM Service Contract

Now that we have covered Architectural, Engineering, and Construction service contracts, let us consider and discuss a sample contract which has been drawn up to satisfy the needs of our hypothetical client, Central City University.

Assuming that C. C. University has reached a tentative oral agreement with its first choice, AECM Consultant Design, Inc., what should happen next? First of all, Central City University should notify the second- and third-choice consultant firms in writing of the University's intent to enter into a contract with Design, Inc. Also, C. C. University should notify Design, Inc. in writing of the intent to enter into a Management for Architectural, Engineering, and Construction Management (MAECM) service contract and request Design, Inc.'s cooperation in the preparation of this contract. Since the drafting of a contract is essentially the work of lawyers, or at least within their general purview, legal advice may be sought by either or both parties at this time. Usually, AECM consultants have had extensive experience in contract preparation, and they are generally willing to prepare this document for the potential client.

Numerous standard forms of contracts and agreements are available for professional services which can be adapted to MAECM work. If one of these standard forms is modified to accommodate MAECM work, it is usually advisable to have competent legal counsel review or assist in making these modifications, remembering full well that the objective of this work is to create a fair, equitable, and legally binding document. Both parties involved in the MAECM service contract should realize that the attorney or attorneys employed by the other party to the contract normally are concerned only with those matters which pertain to their employer's interests.

In any contract for professional services, it is important that a clear understanding be reached regarding relationships between and obligations of all parties, the extents and limits of their work, their responsibilities and authorities, and the compensation each party will receive.

By studying the sample agreement presented in its entirety in this chapter, which has been prepared for MAECM services by Consultant Design, Inc. for C. C. University's project, we will be able to see and comment on what should be included in this type of document. The sample service contract that we shall consider here is, obviously, not an absolutely perfect document. It is, rather, a satisfactory sample document, prepared for illustration and educational purposes. It contains Agreement Cover Sheet, Table of Contents, the Main Body of Agreement Contract, and Appendices A through E.

The items that should be included on the Agreement Cover Sheet (see Figure 7.1) include the title of the document and the names of the principals. Next comes the Table of Contents, which should present in outline form all of the major elements of the agreement, including any appendices, exhibits, or schedules which are attached to and made part of the agreement.

Following the Table of Contents is the sample draft of all Articles, or the main body of the service contract, plus appended exhibits and schedules which are made a part of the contract.

Through reference to this draft sample, either potential clients or consultants may be able to abstract ideas of what should be included in this type of contract. This is only a sample; numerous other forms of service contracts are available which could also be adapted to this type of work, but, in the author's opinion, this sample should serve well as an educational tool and valuable reference.

CONTRACT
between
"CENTRAL CITY UNIVERSITY"
and
AECMM CONSULTANT FIRM
"DESIGN, INC."

New Campus Construction
&
Related Facilities

Figure 7.1. Sample service contract for a MAECM Project. (See pages immediately following for details of the service contract and its appendices.)

TECHNICAL ASSISTANCE SERVICE CONTRACT OUTLINE

Article

Introduction to Contract

1.0　General Terms and Conditions

2.0　Special Conditions

　　2.1.　Description of Project

　　2.2.　Services to Be Rendered by Consultant

　　　　2.2.1.　Construction Management & Project Management Services

　　　　2.2.2.　Architectural & Engineering Services, Studies, & Evaluations

　　　　2.2.3.　Interior Design and Decorating Services

　　　　2.2.4.　Services Concerning Project Equipment

　　　　2.2.5.　Construction Services

　　　　2.2.6.　Services Related to Budgets, Cost Estimates, & Cost Controls

　　　　　　2.2.6.1.　Budgets

　　　　　　2.2.6.2.　Cost Estimates

　　　　　　2.2.6.3.　Cost Controls

　　　　2.2.7.　Miscellaneous Services

　　2.3.　Assignments and Subcontracts

　　2.4.　Data and Services Provided by Owner

　　　　2.4.1.　Site Provision

　　　　2.4.2.　Educational Criteria, Planning Data, Feasibility Studies & Special Analyses

　　　　2.4.3.　Authorized Agent

　　　　2.4.4.　Contractual Commitments

　　　　2.4.5.　Construction Facilities and Equipment

　　　　2.4.6.　Legal Services

3.0.　Consultant's Compensation

　　3.1.　Chargeable Costs

　　3.2.　Guaranteed Maximum Compensation

　　　　3.2.1.　Estimated Total Project Cost

　　　　3.2.2.　Determination of Guaranteed Maximum Compensation

　　　　3.2.3.　Adjustments to Guaranteed Maximum Compensation

　　　　3.2.4.　Division of Savings at Completion

　　　　3.2.5.　Division of Savings on Termination

4.0.　Effective Date of Agreement

5.0.　Entire Contract

Appendices
A. Services to be Rendered by Consultant
   A. Master Planning
   B. Coordination and Supplemental Planning
      1. Coordination
      2. Supplemental Planning
   C. Campus Facilities Design
      1. Conceptual Design
      2. Preliminary Design
      3. Final Design
      4. Contract Documents
      5. Cost Estimates and Budget Estimates
   D. Construction Management
      1. General
      2. Contract Packages
      3. Selection of Construction Contractors
      4. Management of Construction
         a. Administration
         b. Inspection
         c. Miscellaneous Services
   E. Information and Data Provided by Owner
   F. Schedule
B. Chargeable Costs
   A. Total Salary Costs
   B. Non-Salary Costs
      1. Living, Travel, and Other Expenses
      2. Subcontracted Services
      3. Reproducing and Duplicating Services
      4. Communications
      5. Computer Services
      6. Equipment and Materials
      7. Taxes and Licenses
      8. Miscellaneous Expenses
      9. Consultant-Performed Construction Work
C. Payments and Accounting Procedures
   A. Monthly Statement for Compensation
   B. Payments to Consultant
   C. Acceptance and Final Payment
D. General Terms and Conditions
   A. Definitions and General Terms
   B. Independent Consultant

C. Control and Progress of the Work
    1.   Compliance with Laws, Design Standards & Criteria
    2.   Plans and Principles to be Observed
    3.   Subcontracting or Assigning of Contract
    4.   Information and/or Designs & Plans Furnished by Owner
    5.   Inspection
    6.   Ownership of Designs and Plans
    7.   Designation of Representatives
    8.   Nonliability of Owner's Employees
D. Prosecution of the Work
    1.   Progress Reports
    2.   Procedures and Prosecution of Work
E. Alterations, Enlargements & Revisions
F. Suspension or Termination of the Contract
    1.   Force Majeure Suspension
    2.   Payments upon Suspension
    3.   Payments upon Termination
G. General Indemnification
H. Warranty Against Contingent Fees and Gratuities
I. Warranty Against Employment of Owner's Employees
J. Work Week
K. Audit of Direct Costs
L. Responsibilities and Limitations
    1.   Data and Services Furnished by Owner
    2.   Data and Services Furnished by Consultant
    3.   Equipment and Materials
    4.   Construction and Workmanship
    5.   Consequential Damages
M. Patent Rights
N. Notices
E. Insurance
    A. Coverage Required of a Consultant
        1.   Workmen's Compensation and Employer's Liability
        2.   Comprehensive Bodily Injury Liability and Property Damage Liability
        3. "All Risk" Contractor's Equipment Insurance
    B. Certification Required; Notice of Termination
    C. Endorsements or Clauses Required
    D. Coverage Required of Construction Contractors
    E. Coverage Required of Owner

TECHNICAL ASSISTANCE SERVICE CONTRACT

This CONTRACT, entered into this_____ day of_____, 19_____
effective as of the_____day of_____, 19_____, by and between
Central City University, by its president, hereinafter called the "OWNER,"
and AECMM CONSULTANT FIRM, Design, Inc., whose business and post
office address is 10 Main Street, Central City, California, 99999, hereinafter
call the "CONSULTANT,"

WITNESSETH

WHEREAS, the Owner plans a substantial project of development, construction and improvements on property in Central City at the new university 80 acre site, referred to as Parcel 19365 on the Sierra County Assessor's
maps hereinbelow more particularly described and generally referred to
hereinafter as the "PROJECT"; and

WHEREAS, the Owner desires to engage the Consultant to render the
services hereinafter provided for in connection with the Project (hereinafter
defined), as modified or enlarged by the Owner from time to time; and

WHEREAS, the Owner desires the Consultant to provide, directly or
through others, professional services, advice, general administration, and
management services relative to the planning, designing, construction,
furnishing, and equipping of the Project and in all matters incidental thereto;

NOW, THEREFORE, Owner and Consultant, each in consideration of the
undertakings, promises, and agreements on the part of the other herein
contained hereby mutually agree as follows:

Article 1.0—General Terms and Conditions

The "General Terms and Conditions for Consultant Services," hereinafter
called the "General Terms & Conditions," are attached hereto in Appendix
D and are made a part of this contract. In case of conflict between the other
provisions of this contract and said General Terms and Conditions, the other
provisions shall govern.

Article 2.0—Special Conditions

2.1 DESCRIPTION OF PROJECT

The project is a college campus development on property at Central
City, Sierra County, California, consisting of grading, paving, drainage,

utilities, class and administrative buildings, associated service areas, swimming pool, and ancillary facilities. The Project includes Owner modification and/or enlargements as may be made from time to time. Appendix A attached hereto contains a scope of work for the Project.

2.2 SERVICES TO BE RENDERED BY CONSULTANT

The Consultant agrees to provide Engineering, Architectural, Construction Management, and Project Management services for the Project, such services to consist of performing and/or being responsible for the performance of the following services:

2.2.1 Construction Management & Project Management Services

Consultant shall provide all Project Management and Construction Management services, including:

(a) Planning.
(b) Managing the work of:
    (1) Contractors and consultants employed by the Owner.
    (2) Subcontractors employed by the Consultant.
(c) Quality control.
(d) Monitoring costs and providing advice and assistance to the Owner in maintaining control of the costs.
(e) Scheduling.
(f) Coordinating and administering the entire Project work effort.

2.2.2. Architectural & Engineering Services, Studies & Evaluations

Consultant shall provide all architectural and engineering work, studies, evaluations, and plans for the Project, including:

2.2.2.1 Preparation of schematic plans and elevations of the Project, giving due consideration to the Owner's space and functional requirements and directives. Such schematic designs shall serve as a basis for the preliminary estimate of capital requirements to be prepared by the Consultant. When such plans and resultant Project budget are approved by the Owner, they shall serve as general guides to be adhered to in the subsequent work for the Project.

2.2.2.2 Provision and/or direction of professional architectural and engineering services consisting of necessary conferences, preparation of preliminary studies, working drawings, specifications, bills of materials, furnishing all plans and detailed drawings for architectural, structural, plumbing, heating, electrical, mechanical, and other work required for the efficient and proper procurement and installation of materials and equipment and construction of the Project.

2.2.3 Interior Design and Decorating Services

Consultant shall provide professional decorating services with respect to decorating and interior design of the Project, including:

2.2.3.1 Preparing furniture, fixture and equipment layout drawings for dormitories, laboratories, offices, public rooms, restaurants, and service areas.

2.2.3.2 Designing and preparation of specifications for and/or selection and establishment of quantities for:
(a) Furniture, including tables, chairs, beds, chests, desks, sofas, and book stands.
(b) Fixtures, including counters, room dividers, and permanently fixed or built-up laboratory equipment.
(c) Floor coverings, including carpets, rugs, and mats.
(d) Blinds and/or shades, screens and awnings, draperies, and curtains.
(e) Special pictures, paintings, or murals.
(f) Special utility services or special light fixtures and lamps.
(g) Signs, internal and external.

2.2.3.3. Preparation of color schemes and colored renderings for dormitories, libraries, offices, classrooms, and public areas.

2.2.3.4 Supervision of installation of equipment and furnishings.

2.2.4 Services Concerning Project Equipment

Consultant shall establish quantities, design and prepare specifications for, and/or select all equipment to be procured for and installed in the Project.

Consultant shall provide recommendations, assistance, advice, or supervision to Owner relative to the procurement of new, or transfer of existing equipment from the old campus to the Project site.

2.2.5 Construction Services

Consultant shall provide or obtain:

2.2.5.1 A complete boundary and topographic survey of the site, including reference elevations of adjoining properties; together with adequate information on all the rights, restrictions, and easements; and full information on all existing utilities including sanitary and storm sewers, water, gas, and electric lines and services.

2.2.5.2 Soils and subsurface investigations, including all required magnetometer surveys, test pits, test drilling or boring and all chemical, mechanical, or other tests required or advisable.

2.2.5.3 Prepare the necessary documents to obtain bids for procurement by the Owner of construction, labor, materials, and services. This may include the preparation of bid invitations, bid proposal forms for contracts for constructing and equipping the Project, recommended lists of bidders from which bids may be solicited if public bidding is not possible, analysis and evaluation of contractor's proposals, negotiation of terms and provisions of contracts if required, and making recommendations for award.

2.2.5.4 Management and coordination of the work by the construction contractors to accomplish the expeditious completion of the Project. This shall include:

    (a) General administration and management coverage of the project construction, equipment installation, furnishing, and decorating.

    (b) Development of plans, procedures, schedules, records, correspondence, and arrangements for carrying out the construction operations.

    (c) Continuous management and inspection at the site of the Project during all construction. This inspection and management service is to be separate from the day-to-day supervision and other management provided by the contractors who are engaged in constructing and equipping the Project and under contract for such work. Inspecting can include all testing.

    (d) Negotiation, preparation, and recommendation for execution by the Owner's representative *(individual(s) named by Owner)*, for all change orders and all extra work orders necessary or desirable.

    (e) Certification, recommendation, or approval of contractors' invoices for payment.

    (f) Provision of all advisory assistance as required on labor relations matters, safety matters, and on the development of an adequate insurance program.

    (g) On behalf and for the account of the Owner, arrange for special construction services such as hauling, warehousing, plant protection, temporary facilities and utilities, to the extent that such arrangements are not part of the construction contractors' responsibilities and can advantageously be directly provided for outside of existing construction contracts.

2.2.6 Services Related to Budgets, Cost Estimates, & Cost Controls

2.2.6.1 BUDGETS

In cooperation with the Owner, and early in the conceptual stages of design, the Consultant shall establish a price range for the Project's development. This establishment of price range may require the establishment of separate price ranges for numerous layouts and concepts programmed.

The Consultant shall also prepare an original approved Project budget based on an Owner-accepted specific program, layouts, and design information.

When designs, drawings, and specifications have been sufficiently developed, the Consultant shall then prepare a capital budget for the Project.

2.2.6.2 COST ESTIMATES

The Consultant shall prepare, during the course of the Project, such cost estimates of the various components of the Project as he may find necessary in selection between alternates.

Consultant shall prepare an Order of Magnitude Estimate (of construction costs excluding design costs) at the completion of the conceptual design phase for the facilities, and shall prepare Final and/or Advertising Estimates and an Engineer's Estimate for each construction contract. These estimates will be submitted to the Owner at the times designated below:
(a) Order of Magnitude Estimate at completion of conceptual design.
(b) Final Estimate and/or Advertising Estimate at completion of design or prior to contract advertisement.
(c) Engineer's Estimate one or two days prior to receiving bid proposals.

2.2.6.3 COST CONTROLS

The Consultant shall provide a cost control system designed to permit the exercise of continuous control over the costs of the entire Project. This control system shall include, generally, the following services:
(a) Preparation, to the extent not previously provided, periodic estimates of the cost of design, and construction of the project for control purposes. This may include statements of commitments and costs compared to budgets and current forecasts of and to completion.
(b) Preparation of cost reports, progress reports, schedules, esti-

mates of monthly cash requirements, estimates for contract progress payments, and such other reports as requested by the Owner.

(c) Review of invoices and related documents, submitted by construction contractors and material and equipment suppliers, for compliance with contract requirements, approving such invoices for payment and assisting the Owner in allocating contractors' and suppliers' charges in accordance with the Owner's cost accounting system.

(d) Preparation of a "finding of fact" sheet regarding status of work at any time a revised drawing is issued, or any other time a condition develops that, in the judgment of the Consultant, could possibly result in a claim for extra compensation or a time extension to a construction contract.

(e) Review and submittal of recommendations to the Owner regarding the validity of all construction contractors' claims for extra monies or time.

(f) Preparation of estimates of all field change orders and keeping of all records of change order work or work performed on a force account basis.

2.2.7 Miscellaneous Services

The Consultant shall provide the following services:

2.2.7.1 Provide surveying services for the establishment of those lines and grades necessary for the contractors to lay out the work and for periodic checking of the work.

2.2.7.2 Operate a materials testing laboratory or otherwise secure adequate laboratory testing for the testing and control of construction materials and methods.

2.2.7.3 Provide the service of the Consultant's personnel to assist the Owner's operating or maintenance organization in starting and testing the completed work or portions thereof during a trial or preliminary period.

2.2.7.4 Arrange for the delivery to the Owner of "as built" drawings of the completed work, including manufacturers' drawings of major equipment.

2.2.7.5 Furnish to the Owner a "maintenance manual" which gives instructions on maintenance of major or complex equipment, together with providing repair parts lists and similar information.

2.2.7.6 Provide other construction management, administration, procurement, and accounting services as required by the Owner, including the training of key Owner personnel.

2.3 ASSIGNMENTS AND SUBCONTRACTS

2.3.1 To the extent consistent with the effective performance of the work, Consultant may subcontract, with written approval of the Owner, all or part of the engineering or architectural services or any of the work it is obligated to perform hereunder to a third party. Such third party shall not, without written approval of the Owner, subcontract any of the work it is obligated to perform.

2.3.2 It is agreed that, at the election of Consultant, the Consultant and any third party under subcontract with the Consultant shall have the right to have their names appear on all drawings, renderings, and perspectives, and shall receive full professional recognition for all work done hereunder.

2.3.3 This Contract shall not be assignable by either of the parties without the written consent of the other, except as provided in Section 2.3.1, above, and except that it may be assigned without such consent to any corporation with which the party assigning merges or is consolidated, or is affiliated, except that the assignor shall remain liable for the due fulfillment of its obligation under this contract.

2.4 DATA AND SERVICES PROVIDED BY OWNER

2.4.1 Site Provision

The Owner shall provide, without cost to Consultant, a site with right of ingress and egress on which the Project will be constructed.

2.4.2 Educational Criteria, Planning Data, Feasibility Studies, and Special Analyses

The Owner shall endeavor to provide to the Consultant, in accordance with Consultant's project schedule, all preliminary development and operational planning, functional criteria, feasibility studies, and other information as required by Consultant. If such planning, criteria, studies, or other information is not available, the Owner shall authorize the Consultant to perform or obtain it. Compensation for this work will be in accordance with Article 3.0—"Consultant's Compensation" of this contract, except that compensation for this work will not be considered in computing the Guaranteed Maximum Compensation (Section 3.2) to the Consultant.

The Consultant shall be authorized direct contact with the firms supplying or furnishing such information to the Owner, under a separate agreement with the Owner.

2.4.3 Authorized Agent

The Owner shall designate an authorized agent, qualified to make decisions within the terms of the Contract, who will speak for the Owner on all matters pertaining to the prosecution of the work. Owner may also establish a representative to be located in Consultant's office.

2.4.4 Contractual Commitments

Owner shall expeditiously, and in accordance with Consultant's project schedule, perform all acts necessary:
(a) To procure machinery, equipment, materials, and supplies recommended by Consultant pursuant to Section 2.2.4.
(b) To contract for construction services recommended by Consultant pursuant to Section 2.2.5.3.
(c) To contract for consulting services for:
   (1) Preliminary development and operational planning.
   (2) Functional criteria and feasibility studies.
   (3) Other information as required by Consultant pursuant to Section 2.4.2 hereof.

2.4.5 Construction Facilities and Equipment

Owner shall provide to Consultant, or authorize Consultant to provide, at Owner's cost and direction and in accordance with Consultant's project schedule:
(a) Suitable construction field office and test laboratory space, complete with utilities for Consultant's construction management operations.
(b) Suitable warehouse and/or secured area for receiving and storing procured machinery, equipment, materials, and supplies.
(c) Other construction facilities and equipment as required by Consultant.

Costs for aforementioned facilities and equipment shall not be charged against the Guaranteed Maximum Compensation described in Section 3.2 of this contract.

2.4.6 Legal Services

All legal services required by the Project on behalf of the Owner shall be performed by the Owner's counsel or his designated representative; special advice or assistance may be obtained from other counsel by the Consultant for the Owner in matters of contract interpretation.

Article 3.0—Consultant's Compensation

In full compensation to the Consultant for work performed by the Consultant on and after *(effective date)*, with respect to the Project, the Owner shall pay Consultant the amounts provided for in the following sections of this Article 3.0

3.1 CHARGEABLE COSTS

Consultant's compensation shall be the total of the sums provided for under Section 3.1.1, Section 3.1.2, and Section 3.1.3, which follow.

3.1.1 Category 1—Chargeable Costs

Actual cost of base wages and salaries of Consultant's personnel, as indicated in Item A-1 of Appendix B attached hereto, plus twenty-two and one-half percent (22.5%) of such salaries and wages to cover the costs described in Item A-2 of Appendix B; the sum of A-1 and A-2, multiplied by 2.5.[1]

3.1.2 Category 2—Chargeable Costs

Consultant's non-salary costs as described in Items B1-B6, and B8, inclusive, of Appendix B, attached hereto, multiplied by 1.15.

3.1.3 Category 3—Chargeable Costs

Consultant's non-salary costs as described in Items B7 and B9 of Appendix B, attached hereto, multiplied by 1.00.

3.1.4 No charge shall be made under Article 3.0 for the following:

3.1.4.1 Rental or other expenses paid by Consultant for general or project design office space.

3.1.4.2 Lighting, power, office furniture, or janitorial services.

Payment by the Owner to Consultant for such Chargeable Costs shall be made monthly during the term of this contract, in accordance with the provisions of Appendix E, attached hereto.

3.2 GUARANTEED MAXIMUM COMPENSATION

Notwithstanding the provisions of Section 3.1, above, the maximum total compensation to be paid to Consultant for the work, as set forth in

---

[1]*This multiplier is a negotiable variable which generally lies between 1.5 and 3.0.*

Article 2.0 and Appendix A of the contract, shall not exceed the Guaranteed Maximum Compensation as determined under the provisions of this Section 3.2. Such Guaranteed Maximum Compensation may be mutually adjusted from time to time, as hereinafter provided.[2]

3.2.1 Estimated Total Project Cost

In addition to the other budgets, estimates, and cost controls for the Project, the Consultant shall prepare an Estimate of Total Project Costs. This estimate is to be prepared when, in the opinion of the Consultant, the costs of the Project can be adequately defined, estimated, and/or projected. At this time, Consultant shall prepare and deliver to the Owner an Estimated Total Project Cost which shall include a definition of the scope of the work included in the estimate, together with an estimate and/or projection of the total construction cost to the Owner, including, but not limited to, the following:

(a) Cost (or in the case of used equipment which the Owner already owns and elects to transfer to this Project, either the market value or the worth from generally accepted accounting principles) of machinery, equipment and supplies, including demurrage or freight thereon, loading, unloading, warehousing, storage, handling, and all other related costs.

(b) Costs of construction, including all commitments and/or payments for labor and services and to construction contractors and subcontractors, but excluding land costs.

(c) Costs of Consultant services deemed to be extras to this contract.

(d) Cost of other consultants employed by the Owner, if any.

(e) Compensation to Consultant, pursuant to Section 3.1.

(f) Sales taxes, use taxes, and general excise taxes (excluding, however, taxes measured by net income of Consultant).

(g) Escalation of the estimated or projected cost of materials, labor, and equipment or other expenses to the completion of the Project.

(h) A contingency factor.

The Estimate of Total Project Cost need not be broken down as de-

---

[2]*The Guaranteed Maximum Compensation clauses of a contract are usually negotiable. The basis of this Guaranteed Maximum Compensation is usually one of the following: (1) construction costs or estimated construction costs, or (2) construction costs or estimated costs plus AECMM costs times a multiplier that should range between 7 and 17 percent, dependent on the project size and complexity. For this example, the basis for establishing the Guaranteed Maximum Compensation shall be the "estimated total project cost," which includes the estimated construction costs; the estimated cost of work design only; and Architectural, Engineering, and Construction Management and Project Management fees, together with other Project related costs.*

tailed in Items (a) through (h) above, but the estimated cost of each of the major (5 percent or more of the Estimate of Total Project Cost) components of the Project shall be separately stated, and the criteria upon which these costs are based shall be identified.

3.2.2 Determination of Guaranteed Maximum Compensation

The Guaranteed Maximum Compensation to Consultant shall be an amount equal to eleven percent (11%) of the Estimated Total Project Cost, established pursuant to Section 3.2.1 above, except for work designed only. For purposes of changed work and for work designed only, this portion of the Guaranteed Maximum Compensation section shall be considered to consist of design services at eight percent (8%) and construction services at three percent (3%) of the estimated construction cost of such work, unless with respect to any particular change such breakdown would be inequitable, in which case an appropriate percentage will be negotiated.

The compensation to Consultant for changed work for work designed only shall not be held separately, but shall be included in the Estimated Total Project Cost, to which the eleven percent (11%) multiplier shall be applied.

3.2.3 Adjustments to Guaranteed Maximum Compensation

The Guaranteed Maximum Compensation, determined pursuant to Section 3.2.2 above, shall be adjusted to reflect each of the following matters:

3.2.3.1 The mutually agreed upon adjustments relating to each alteration or enlargement of the Project or of Consultant's work directed by the Owner, determined pursuant to the provisions of Item E-1 of Appendix D, attached hereto.

3.2.3.2 The mutually agreed upon adjustments or revisions of the Estimated Total Project Cost or to the criteria supporting this estimate, resulting from Owner-directed revisions, pursuant to Item E-2 of Appendix D, attached hereto, or based on mutually acceptable criteria.

3.2.3.3 Costs determined, pursuant to Section 3.1 hereof, resulting from:
(a) Force majeure.
(b) Any delays to Consultant's work or to Consultant's schedule of Project completion caused primarily by acts of the Owner or failure of Owner to make decisions or commitments in accordance with the Contract or the Appendices attached thereto, except that no adjustment to Guaranteed Maximum Compensation shall be allowed beyond the twentieth day for these

causes if the Consultant does not advise the Owner in writing of any such delay prior to the twentieth day of such delay.

(c) Suspension of work as directed by Owner, pursuant to Item F-1 of Appendix D.

3.2.3.4 The cost, determined in accordance with Section 3.1 hereof, of any additional or extra work performed by Consultant. This includes work caused by failure of Owner to provide any of the data or services or failure to perform acts which the Owner is obligated herein to provide or perform, especially as described in Section 2.4 hereto.

3.2.3.5 The cost of work performed by Consultant which was rendered useless because of Owner alterations, prior to the establishment of the Estimated Total Project Cost. The estimated construction cost of this work may be included in Guaranteed Maximum Compensation computations under the category of "work designed only," to which a mutually agreed upon multiplier of eight percent (8%) or less shall be applied for determination of this portion of the Guaranteed Maximum Compensation.

3.2.3.6 The cost of termination of the Contract, pursuant to Item F-2 of Appendix D hereto, shall not be included in the Estimated Total Project Cost.

3.2.4 Division of Savings at Completion

In the event Consultant compensation, pursuant to Section 3.1, at the completion of work is less than the adjusted Guaranteed Maximum Compensation, pursuant to Sections 3.2.2 and 3.2.3, the resulting difference, hereinafter called "Savings," shall be divided between the Owner and Consultant. The Owner shall retain seventy-five percent (75%) of such Savings, and shall pay to Consultant the remaining twenty-five percent (25%) of such Savings.

3.2.5 Division of Savings on Termination

In the event of Contract termination, a proration of the estimated contract Savings shall be made in accordance with the percentage of Architectural, Engineering, Construction Management and Project Management services which are complete at the time of termination, based on a projected or estimated service billing and the Estimated Total Project Estimate. The prorated Savings, which shall exclude termination costs, shall be divided equally between the Owner and the Consultant; the Owner shall pay to the Consultant fifty percent (50%) of this prorated Savings.

Article 4.0—Effective Date of Agreement

The Contract is effective as of the _____ day of _____ , 19 _____

Article 5.0—Entire Contract

This Contract and the Appendices attached hereto and listed below consti-tute the entire Contract between the parties and supersedes all previous agreements or understandings with respect to the subject matter hereof. Consultant's duties, obligations, and liabilities hereunder shall be limited to those expressly provided in the Contract and the Appendices thereto, and no other duties, obligations, liabilities, or warranties shall be implied.

Attachments

The following attachments are appended to and made a part of this Contract:

    Appendix A: Services to be Rendered by AECMM Team (Consultant)

    Appendix B: Chargeable Costs

    Appendix C: Payments and Accounting Procedures

    Appendix D: General Terms and Conditions

    Appendix E: Insurance

This Contract may be changed, but only by a Contract Modification signed by the parties hereto.

IN WITNESS WHEREOF, the parties hereto affix their signatures to this Contract as of the day, month and year hereinafter written.

    CENTRAL CITY UNIVERSITY (OWNER)

            By: _____Date:_____
AECMM CONSULTANT FIRM DESIGN, INC. (CONSULTANT)
            By: _____Date:_____

Appendix A

Services to be Rendered by Consultant
(AECMM CONSULTANT FIRM DESIGN, INC.)

Consultant shall provide all of the Architectural, Engineering, Construction Management and Project Management services required in connection with the development of the new campus facilities for Owner. The initial scope of these services shall consist of planning, design, management, and construction management for the first phase of the new campus development as defined in the material or approved supplemental material of the Owner-approved Master Development Plan for the new facilities.

These services shall include, but not be limited to, the following:

A. MASTER PLANNING

Conduct master and supplemental planning studies, including determination of the most feasible location and alignment of facilities, and preparation of a master facilities layout of the new campus site and facilities.

The Consultant shall, in the preparation of the campus layout plan, include the following services:

1. Surveying and mapping.
2. Determination of facilities boundaries for the ultimate boundaries and possible phase developments.
3. Preparation of campus layout plans and location maps for utility systems.

B. COORDINATION AND SUPPLEMENTAL PLANNING

1. Coordination

    a. Consultant shall coordinate all programs and schedules that are required and authorized by the Owner for the development of the new campus facility. This may include programs and schedules of contractors, engineers, architects, consultants, and others who are authorized by the Owner to participate in the Project. Consultant shall coordinate the scheduling of each phase and item of work and shall be authorized to guide the work in such a manner that it will conform to the established schedules.
    b. The Owner will establish a University Facilities Users Committee which may be comprised of representatives of the faculty, student body, concessionaires, government agencies, and other interested groups. The Consultant shall schedule and

conduct meetings of this committee as required to coordinate the requirements of the campus facilities users.

2. Supplemental Planning

Consultant shall conduct planning studies as required to supplement the approved Master Development Plan, which may include the following:

    a. Reviews and preparation of campus development plans with regard to traffic and utilities.

    b. Establishment of road and building configurations and locations for the new campus.

    c. Planning for maintenance facilities.

    d. Planning for utility systems.

    e. Planning of transportation and circulation systems.

    f. Planning for faculty and student housing, including providing space requirement criteria and growth projections.

    g. Planning motor vehicular access and parking facilities.

C. CAMPUS FACILITIES DESIGN

Consultant shall perform all design services for the new campus facilities, as follows:

1. Conceptual Design

Consultant shall perform all architectural and engineering services required to develop a conceptual design for the complete campus facilities as required by the approved Master Development Plan until the year 19____ . The conceptual design shall be adequate to proceed without delay into preliminary and final design, and shall provide information or design criteria on the following subjects or systems:

    a. The topographic and geological conditions of the site.

    b. All transportation and circulation, including vehicle traffic and parking facilities.

    c. Expected drainage and soils conditions.

    d. Dormitories, houses, and residential facilities which are part of the campus development, or those outside the development which should be considered.

    e. Administration, tenant, and vendor facilities.

    f. Utility systems.

    g. Campus operational requirements and support services.

2. Preliminary Design

Consultant shall prepare preliminary plans and outline specifications for the first stage construction of the approved campus concept including the campus administrative, classroom, and residence buildings; access roads; and parking facilities. Consultant shall also conduct site investigations including geological surveys, test boring programs, soil sampling and testing, and other investigations as necessary.

a. The preliminary plans shall include sufficient information to delineate and define the work.

b. The preliminary drawings shall contain sufficient detail to inform the campus users and administrators of the facilities to be provided. Preliminary design drawings shall include:

    (1) General arrangement drawings for the overall facility and for roadway system, parking facilities, and buildings.

    (2) Architectural plans, elevations, and sections for all major buildings.

    (3) Double-line drawings of developed floor plans.

    (4) Functional layouts and single-line diagrams for all electrical and mechanical systems.

    (5) Plans showing improvements to existing topographic features, locations and dimensions of parking areas, access roads, driveway walks, landscaped areas, etc.

c. Outline specifications shall be prepared in sufficient detail to illustrate the type, character, and quality of the structures, construction materials, and equipment to be installed.

d. Sufficient data, data analysis, and background shall be prepared with the preliminary plans and specifications to explain adequately and justify the proposed design.

e. The design criteria shall be assembled, which shall serve as a basis for the final design of the facility, including buildings, mechanical, electrical, civil, and structural systems.

3. Final Design

Final design shall include detailed plans and specifications suitable for construction of the proposed first phase facilities. Plans shall include location and layout plans; plot plans; profiles; floor plans; elevations; sections; details and all pertinent information necessary for bidding, award of contracts, and construction.

4. Contract Documents

The Consultant shall prepare all documents required for the award of construction contracts for the proposed facilities. The documents for each construction contract may include, but are not limited to, the following:

a. Notice to Bidders.

b. Instructions to Bidders.

   c. Special Provisions.
   d. General Provisions.
   e. Technical Provisions.
   f. Proposal.
   g. Contract Forms.
   h. Contract Drawings.

5. Cost Estimates and Budget Estimates

   During the master planning, and early in the conceptual design phase, Consultant shall provide, as required, Budget Estimates which may include probable ranges of costs for the schemes, layouts, or proposed concepts.

   Consultant shall prepare an Order of Magnitude Estimate for the approved concept at the completion of conceptual design phase for the facility.

   Consultant shall prepare an Advertising or PreFinal Engineer's Estimate for each construction contract, to be submitted to the Owner together with the contract documents for review prior to advertisement for bids which, if the Owner directs, will be handled by the Consultant.

D. CONSTRUCTION MANAGEMENT

1. General

   Consultant shall manage the total construction of all campus facilities, together with any additional related projects assigned to the Consultant by the Owner's Board of Regents.

2. Contract Packages

   In order to assure a timely and economic completion of the project, the Consultant may be required to prepare separate construction contracts for the various portions of the work. The exact scope of each contract package may be adjusted as required by the actual progress of design, budget considerations, or construction activities.

3. Selection of Construction Contractors

   Consultant shall assist the Owner in reviewing and evaluating the bids received for the construction work.
   This assistance shall include:
   a. Examination and analysis of the bids received.
   b. Tabulation of the bids, including a listing of the unit prices and lump sum prices bid and the extension of these prices.
   c. Determination of the qualifications of bidders to perform the proposed work.

    d.   Checking the legality of bids received.
    e.   Negotiations with the contractors on any additional provisions.
    f.   Reporting to the Owner on the feasibility of awarding contracts for construction on the basis of bids received.
    g.   Making a Consultant's recommendation to the Owner.

4. Management of Construction

Consultant shall manage and administer the all new campus facilities and related construction for the Owner. These services shall include the following:

    a.   ADMINISTRATION

       (1) Organizing, planning, and managing the construction program.
       (2) Assisting the Owner to obtain, at Owner's expense, all requisite building permits and other permits and licenses from appropriate government authorities.
       (3) Assisting the Owner as required in contractual matters related to the construction of the Project.
       (4) Providing technical and administrative assistance to the Owner in the preparation, verification, or authorization of contract progress pay estimates.
       (5) Providing advisory assistance on labor relations and safety matters, as requested by the Owner.
       (6) Providing advisory assistance, as requested, in the development of an adequate insurance program for the Owner's protection.
       (7) Preparing cost reports, estimates of monthly cash requirements, and such other cost reports as requested by the Owner.
       (8) Assisting the Owner with engineering, purchasing, or expediting manufacture or delivery of equipment purchased directly by the Owner.
       (9) Performing actual construction work when, in the opinion of the Consultant and with the written approval of the Owner, it is determined that this is the most expeditious way to perform work.

    b.   INSPECTION

       (1) Organizing, staffing, and managing a construction inspection program.
       (2) Coordinating and inspecting the work of the contractors to secure completion and fulfillment of the contracts in accordance with plans, specifications, and approved schedules.

(3) Preparing daily, weekly, and monthly progress reports.

(4) Documenting the activities of the contractors.

(5) Taking progress photographs and documentary photographs of the Project prior to, during, and after completion of construction.

c. MISCELLANEOUS SERVICES

(1) Providing surveying services for the establishment of those lines and grades necessary for building contractors to lay out the building work; providing construction surveying and staking services for engineering contractor's work; periodically checking finished work and performing surveys as necessary to determine the actual quantity of work performed.

(2) Providing materials testing services for the Project. This may include the operation of a materials testing laboratory or facilities at the Project site for testing and control of construction materials and methods.

(3) Arranging for the supply of "as built" drawings of the completed work, including manufacturers' drawings of major equipment.

(4) Providing the services of Consultant's personnel to assist the Owner's operating and maintenance organization during a trial or preliminary operation period.

(5) Providing other project-related administration and accounting services, as requested by Owner.

E. INFORMATION AND DATA PROVIDED BY OWNER

The Owner shall provide or make available to the Consultant all data and information relative to this Project which it has acquired prior to entering into the Contract, or which it acquires after entering the Contract, of which the Owner deems it would like to, or should inform the Consultant.

F. SCHEDULE

Consultant shall schedule and coordinate all work so that each portion of the work is completed in or before the dates specified below:

1. Complete Master Planning and submit report to the owner for review.   January 1, 1978

2. Complete supplemental planning for owner review.   April 1, 1978

3. Submit layout plans and partially complete concepts for review.   June 1, 1978

4. Complete conceptual design and submit Order   August 1, 1978
    of Magnitude Estimate.
5. Completion of preliminary design for early con-   October 1, 1978
    tracts to be constructed during the first phase
    construction.
6. Submit bid documents for first construction con-   January 1, 1979
    tract.
7. Complete construction of minimum operational   September 1, 1980
    facility.
8. Complete first phase construction.   January 1, 1981

The dates by which work is to be completed shown in the above
Schedule are contingent upon prompt review of submitted materials by
all parties concerned. If comments or submitted materials are received
later than fourteen (14) calendar days after submittal, the dates in the
above Schedule shall be adjusted as mutually agreed upon between the
Owner and Consultant.

Appendix B

CHARGEABLE COSTS

For Architectural, Engineering, Construction Management, Project Management, Purchasing, and other services performed by Consultant in accordance with the Contract to which this Appendix B is attached, Consultant shall charge the Owner with the costs incurred by the Consultant as set forth below:

A. TOTAL SALARY COSTS

   Total Salary Costs shall consist of the SUM of the following:
   1. Actual cost of wages and salaries for such time as is exclusively devoted to the work by any of Consultant's employees. These employees can include managers, administrators, constructors, supervisors, engineers, architects, designers, draftsmen, purchasers, expeditors, inspectors, estimators, cost engineers, operators, accountants, stenographers, secretaries, clerks, and others.[3]
   2. A fixed amount of twenty-eight and one-half percent (28.5%) of the actual cost of wages and salaries, as compensation to Consultant for applicable payroll insurance and payroll taxes, holidays, vacations, sick leave, group health, accident and life insurance, public liability, property damage liability, and other insurance customarily maintained by Consultant; and all other employee benefits and welfare expenses, including Unemployment Compensation and Workmens Compensation.[4]

      In the event the aggregate cost to Consultant of the aforesaid payroll insurance, payroll taxes, and other employee benefits and welfare expense costs shall increase after the effective date of the contract to which this Appendix B is appended and made a part, no adjustment shall be made in the foregoing percentage amount (Item 2, above), except by mutual agreement and a Contract Modification for such additional costs.

B. NON-SALARY COSTS

   1. *Living, Travel, and Other Expenses* of Consultant's personnel while traveling away from Consultant's permanent offices in the interest of the work, and if prior written approval is obtained from the Owner.

[3]*Salary schedules can be obtained and included in this section to insure consistent salary practices.*
[4]*The applicable percentage rate is negotiable, and is usually between 20% and 35%.*

Transportation shall be by an economical means consistent with the time available and urgency of the trip. An allowance of $30 per diem to cover all lodging costs, meals, personal telephone calls and telegrams, laundry and valet services, fees, and tips will be provided.

2. *Subcontracted Services*, consultant fees, and associated expenses, including special accounting expenses.

3. *Reproducing and Duplicating* services and materials, such as blueprints, photostats, photographs, Xerox, ditto, mimeograph, printing, and binding.

4. *Communications* such as telephone expenses, telegrams, teletypewriter messages, cablegrams, and radiograms.

5. *Computer Services* performed by Consultant, based on the Consultant's current hourly rates or at actual cost, if performed by subcontract.

6. *Equipment and Materials* purchases, rental and lease costs applicable to the work, subject to applicable laws. Such office equipment as adding machines, typewriters, and calculators, if not available from the Owner, shall be purchased with the provision that, at the end of their use on the Project, they shall be either transferred to the Owner or disposed of as directed by the Owner, with a corresponding credit to Consultant, if any. Permanently assigned automobiles shall be reimbursed at the rate of $135 per month per vehicle.

7. *Taxes and Licenses*, assessments, levies, imposts, duties, and excises, excepting only those taxes levied directly on or measured by Consultant's net income and excepting the cost of licenses or permits required by governmental authorities in order for Consultant to carry on business in the jurisdiction in which the work is performed.

8. *Miscellaneous* and other expenses, as approved by the Owner, which are incurred in direct performance of the work.

9. *Consultant-Performed Construction Work*. In the event the Consultant is required to perform actual construction work, costs shall be determined as established herein, except that a charge in the amount of one hundred fifteen percent (115%) of the actual direct payroll expense shall be made for Labor (members of organized unions, foremen, and superintendents), for materials and equipment or subcontracts made directly with the Consultant.

For purposes of computing Guaranteed Maximum Compensation and Savings, such Consultant-performed construction work shall be considered to have been performed by a separate contractor or subcontractor.

Appendix C

PAYMENTS AND ACCOUNTING PROCEDURES

Payments by the Owner to the Consultant under the Contract to which this Appendix C is attached shall be made as hereinafter provided.

A. MONTHLY STATEMENT FOR COMPENSATION

A Statement for Compensation shall be rendered by Consultant monthly and upon completion of the work. Such Statement for Compensation shall be based on Consultant's accounting system, which shall be closed monthly. The amount of payment requested on any monthly Statement for Compensation shall include all payments due Consultant under the Contract as a result of work performed during such monthly accounting period, or as a result of work performed in previous monthly accounting periods and not theretofore invoiced to the Owner. Consultant shall submit with each monthly Statement for Compensation such payroll records, invoices, and other documents satisfactory to the Owner as may be required to support such statement.

B. PAYMENTS TO CONSULTANT

Upon approval of each monthly Statement for Compensation by the Owner, payment shall be made to Consultant as soon as is practicable but, in all cases, prior to forty-five (45) days after receipt of such Statement, provided there is no misrepresentation or error in such Statement.

Monthly payments to Consultant shall be in the amount of ninety percent (90%) of the amount shown on such monthly Statements for Compensation, less all previous payments, provided however, that if the president and comptroller of the Owner-University jointly determine that the amount of total retained percentage is in excess of the amount considered by them to be adequate for the protection of the Owner, they may, at their discretion, release to the Consultant such excess amount.

Within thirty (30) days after completion by Consultant of the Consultant's work under this Contract, to which this Appendix C is attached and made a part, (excluding any remedial work under Appendix D, or within thirty (30) days after any termination of the Contract by the Owner), payment shall be made by Owner to Consultant for all amounts due Consultant under the Contract including, without limitation, all amounts withheld pursuant to this section.

C. ACCEPTANCE AND FINAL PAYMENT

When, in the opinion of Consultant, all of the Consultant's work hereunder is complete, or at such time as the Owner's representative may designate in writing to Consultant that, in his opinion, the work hereunder is substantially complete and services including charges to Owner should cease, the Consultant shall notify the Owner of such completion in writing. Within thirty (30) days after receipt of such notice, Owner shall either:

1. Provide its written notice of acceptance to Consultant, or
2. Give Consultant written notice of any unfinished items or deficiencies to be corrected.

    In this event, Consultant shall notify the Owner in writing when all such items of work have been finished or deficiencies corrected, and the Owner shall give its notice of acceptance in writing within thirty (30) days thereafter.

Such final acceptance of the work shall not relieve Consultant from any liability to the Owner, determined pursuant to Appendix D, General Terms and Conditions, resulting from improperly prepared plans or specifications.

Upon final acceptance of the work done by Consultant, the Owner shall pay to the Consultant the unpaid balance of any monies due for the work, including retained percentages, provided that, prior to final payment under the Contract, or prior to settlement upon termination of the Contract, and as a condition precedent thereto, the Consultant shall execute and deliver to the Owner's representative a release from all extra work, changed work, or other cost claims against the Owner, together with lien releases and tax releases or clearance. Further, it is agreed that acceptance by the Consultant of final payment shall constitute payment in full for all work done.

Appendix D

GENERAL TERMS AND CONDITIONS

The General Terms and Conditions applying to the Contract to which this Appendix D is attached are set forth below.

A. DEFINITIONS AND GENERAL TERMS
1. *Owner:* Central City University, located in Central City, California.
2. *Consultant:* AECMM Consultant Design, Inc., whose business and post office address is 10 Main Street, Central City, California, 99999.
3. *Comptroller:* The Comptroller, Central City University, acting either directly or through his duly authorized representative or subordinate.
4. *Subcontractor:* The written agreement between Consultant and any other consultant or separate contractor which sets forth obligations of the parties thereto, together with any mutually agreed amendments or modifications to the initial agreement.
5. *Design and Plans:* Any designs, plans, drawings, specifications, cost estimates, proposed schedules, studies, reports, and other items required under the Contract.
6. *Contract:* The written agreement between Central City University, the Owner, and AECM Consultant Firm Design, Inc., the Consultant, to which this Appendix D is attached, and any mutually agreed to amendments or modifications thereto, setting forth the obligations of the parties.
7. *Final Plans and Specifications:* Final Plans and Specifications shall be the construction specifications and drawings provided by Consultant for a specific construction package.
8. *Project:* As defined in Article 2.1 of the Contract.
9. *Work:* As defined in Contract documents or indicated as "services" or otherwise herein.
10. *General Notes:*
    a. The titles or headings of the Items are intended for convenience of reference, and shall not be considered as having any bearing on their interpretations.
    b. References to all governmental laws, ordinances, codes, rules and regulations shall include all amendments thereto.
    c. All words used in the singular shall extend to and include the plural.
    d. All words used in the plural shall extend to and include the singular.

e. All words used in any gender shall extend to and include all genders.

B. INDEPENDENT CONSULTANT

In the performance of the Contract, Consultant shall act as an independent Consultant, maintaining full and complete control over Consultant's employees.

C. CONTROL AND PROGRESS OF THE WORK

1. Compliance with Laws and Design Standards & Criteria

   a. The Consultant shall familiarize himself with and shall at all times use his best efforts to comply with and observe all applicable Federal, State, County, and municipal laws, ordinances, codes, design standards, rules and regulations, and other criteria which in any manner affect his services. If any discrepancy or inconsistency which pertains to the work is discovered by Consultant in such laws, ordinances, codes, design standards, etc., the Consultant shall immediately report the same to the Owner in writing and concurrently make a definite effort to resolve such problems that the discrepancy might create.

   b. If Consultant discovers any apparent error or omission in the information and/or designs and plans furnished by the Owner, it shall immediately notify the Owner's representatives in writing. The Owner's representatives shall then make such corrections and interpretations as may be deemed necessary for performing the work under the Contract.

2. Plans and Principles to be Observed

   Consultant shall familiarize himself with and shall at all times endeavor to comply with, observe, and conform to the State, County, and Municipal general plans, all applicable approved development plans, setback limitations, right-of-way, established objectives, and principles. If any discrepancy, inconsistency, or conflict is discovered by the Consultant in the course of the work or in plans which relate to the project, Consultant shall report this finding to the Owner in writing. Consultant shall direct his services to relate appropriately to and in accordance with established engineering and/or architectural design principles and practices, and to the natural and man-made environment.

3. Subcontracting or Assigning of Contract

   The Consultant shall not subcontract or assign all or any part of the work without the prior written consent of the Owner, and any consent by the Owner to subcontract, assign, or otherwise dispose of any portion of the work shall not be construed to relieve Consultant of any responsibility for the fulfillment of the work under the Contract.

4. Information and/or Designs & Plans Furnished by Owner

The Owner shall fully furnish information for the guidance of Consultant's work and/or, at the Owner's option, shall provide actual designs and plans for that portion of the Project which it so elects to incorporate into Consultant's work. It is agreed that Consultant will not be responsible for the accuracy and adequacy of such furnished information and/or designs and plans.

5. Inspection

The Owner, through its designated representative, may inspect the work of Consultant and his subcontractors at any time.

6. Ownership of Designs and Plans

Upon expiration or termination of the Contract, all designs and plans shall become the sole property of the Owner. Consultant shall compile and submit in an orderly manner to the Owner all designs and plans prepared by the Consultant in the course of the execution of the work under the Contract. Consultant shall also furnish "as-built" drawings based on marked-up or revised design drawings, showing all modifications, changes, or revisions to the initial contract design drawings furnished to construction contractors. Consultant shall have the right to retain copies of all designs and plans, and to use them in the furtherance of the Consultant firm's business or its own technical competence.

7. Designation of Representatives

   a. The Owner shall designate, in writing, a representative of the University to approve and coordinate the work under this Contract; and to act as principal liaison between the Consultant and Owner, to answer any questions, and to expedite decisions and progress reports.
   b. The Consultant shall designate in writing, with the approval of the Owner, a project representative who will maintain close and frequent communications with the Owner's representatives. The Consultant's project representative shall be experienced and qualified in the type of work involved and shall be directly responsible for the prosecution of the work under the Contract.

  c. Every effort will be made by all parties to the Contract to retain the same liaison representatives during the term of the Contract in order to maintain continuity of effort and control.

 8. Nonliability of Owner's Employees

 The President of the University and any of the Owner's duly authorized representatives and subordinates, in carrying out the provisions of the Contract or in exercising any power or authority granted herein, shall not be held personally liable in any way, it being understood that, in such matters, they act as agents and representatives of the Owner.

D. PROSECUTION OF THE WORK

 1. Progress Reports

  a. Consultant shall be available upon reasonable demand to discuss the progress of the work.

  b. Unless otherwise instructed in writing by the Owner, Consultant shall submit to the Owner, at the end of each calendar month, a written narrative progress report of the work being performed and the approximate percentage, by identifiable phases and/or items, of work that is complete. This report shall also indicate the approximate overall percentage of the Consultant's services that are complete, together with a cumulative total of all billings submitted by the Consultant to the Owner.

 2. Procedures and Prosecution of Work

  a. Unless otherwise authorized in writing by the Owner, all designs and plans to be submitted by Consultant to the Owner for review, approval, or record shall be drawn directly or reproduced on 30″ x 40″ 1000-H tracing or equivalent paper.

  b. Each sheet of the plans shall be endorsed by the Consultant and/or the appropriate registered architect, engineer, or surveyor in the manner prescribed by law. Where the work, or a subdivision of the work, involves only landscaping, and if the laws provide, the landscape architect who prepared or supervised the preparation of the plans or of the sheet shall endorse it.

  c. All plans and specifications shall be submitted by Consultant to the Owner for the Owner's approval in writing. This includes all revisions to plans and all drawing changes made to enlarge, change, or facilitate the construction, except that the Owner may, in writing, authorize the Consultant to act in the interest of the Owner in making such revisions and drawing changes that are required by exigencies of the construction operations. The approval of the plans and specifications by the Owner shall not be construed to relieve the Consultant of the responsibility for correcting any errors or discrepancies in the plans or specifications which may become apparent to the Consultant even after approval has been given, nor shall the approval be construed to relieve the Consultant of the responsibility of using his best efforts for designing the Project to conform to all applicable design standards and criteria, laws, plans, established engineering, architectural, and landscape planning principles and practices.

E. ALTERATIONS, ENLARGEMENTS & REVISIONS

1. Owner reserves the right, from time to time and by written notice to Consultant, to make Changes, including both:

a. Alterations or enlargements of the Project or the Consultant's work, or

b. Revisions to any of the criteria or design data supporting the Estimated Total Project Cost established by Consultant pursuant to Article 3.2.1 of the Contract.

2. Upon issuance by the Owner to the Consultant of such written notice, Consultant shall make an estimate of the effect of the Change on the Project time schedule, together with estimates of the cost of any Change and the cost effect of any Change on the Estimated Total Project Cost and the Guaranteed Maximum Compensation.

3. Consultant shall commence work on such Change upon:

a. Written instructions from the Owner to proceed, and

b. Owner approval, in writing, of Consultant's estimated effect on Project time, and

c. Owner and Consultant agreement on the amount of adjustment (on account of such Change) to the Estimated Total Project Cost and the Guaranteed Maximum Compensation, determined in accordance with Article 3.2.3 of the Contract.

F. SUSPENSION OR TERMINATION OF THE CONTRACT

Owner shall have the right to suspend the work or terminate this Contract at any time by giving written notice to Consultant in sufficient time before suspension or termination to allow for an orderly shutdown of Consultant's project service operations. Upon receipt of any such notice, Consultant shall suspend or terminate all work under the Contract as quickly as practicable. In the case of termination, the termination period shall not exceed sixty (60) days without written consent of the Owner.

1. Force Majeure Suspension

   The duties and obligations of the parties to this contract shall be suspended during such time as performance by either party is prevented or materially impeded by strikes, labor disturbances, riots, fire, government act, war, acts of God, or any other causes similar to the foregoing and beyond the control of the parties hereto. No such suspension, however, shall suspend, alter, or affect the Consultant's right to receive payments for chargeable costs of the work.

2. Payments upon Suspension

   In the event of suspension of work by the Owner or Force Majeure suspension of the work as provided for above, the Owner shall pay to the Consultant, upon receipt of invoice therefore, all costs determined as provided in Article 3.1 of the Contract incurred by Consultant during, or as a result of each suspension. Such payment shall include, but not be limited to:

   a. Costs of maintaining Consultant's staff of architects, engineers, and construction management personnel and supporting personnel available in sufficient strength and at the proper location to permit prompt resumption and efficient prosecution of the work upon the cessation of the suspension.

   b. Costs of relocating personnel, including relocation expenses of employees' dependents, to other work; this cost is limited to $2,500 per employee relocated.

   c. Costs of recruitment of personnel to replace personnel terminated or lost through other attrition during, or as a result of such suspension.

   d. Costs incurred in respect to subcontracts issued by the Consultant pursuant to Article 2.3 of the Contract, together with an amount to be mutually agreed upon by Owner and Consultant to compensate Consultant for the costs associated with the suspension.

3. Payments upon Termination

   In the event of termination of the Contract by Owner, the Owner shall pay to Consultant, upon receipt of invoices therefor:

   a. All chargeable costs payable to Consultant pursuant to Article 3.0 of the Contract, in respect of work performed by Consultant to the effective date of termination minus all payments of such chargeable costs previously paid to Consultant by Owner.

b. All costs, determined as provided in Article 3.1 of the Contract, incurred by Consultant for:

(1) Relocation of Consultant's personnel, including relocation expenses of employees' dependents. Such relocation expenses shall be limited to $2,500 per employee so relocated.

(2) Demobilization of its forces; this cost may include a charge not to exceed $2,500 for each employee terminated or lost through other attrition during or as a result of such termination, to compensate Consultant for cost of recruitment of personnel to replace such personnel so terminated or lost through attrition.

(3) Costs of unfulfilled commitments and/or cancellation charges under uncompleted subcontracts issued by Consultant pursuant to Article 2.3 of the Contract.

(4) All other costs incurred by Consultant as a result of the Owner's termination of the Contract prior to completion of the work, including a settlement for any anticipated Savings participation as described in Articles 3.2.4 and 3.2.5 of the Contract which the Consultant can adequately project. The Owner will negotiate for a settlement of anticipated Savings participation on a pro-rata basis, as described in Article 3.2.5 of the Contract, or may elect to make a cash settlement with the Consultant in the amount of an additional twenty percent (20%) of the total compensation due the Consultant after complete termination.

G. GENERAL INDEMNIFICATION

1. Except as otherwise provided in this Contract, Consultant agrees to indemnify, hold harmless, and defend Owner from any and all losses, claims, expenses, liabilities, and judgments for personal injury or death or for damage to property arising out of the negligent operations of the Consultant under the Contract, except for such personal injury or death or property damage as is cause by the sole negligence of the Owner. Consultant's liability under this Item shall be limited to the risks covered by, and the proceeds of the applicable insurance required pursuant to Appendix E, attached hereto.

2. Consultant shall obtain and maintain all insurance as provided in Appendix E, attached hereto. Lack of evidence of such coverage shall be grounds for suspension or termination of this Contract.

H. WARRANTY AGAINST CONTINGENT FEES AND GRATUITIES

1. Consultant warrants that he has not and will not employ or retain any company or person, other than a bona fide employee working solely for Consultant, to solicit or secure the Contract, and that it has not paid or agreed to pay any company or person, other than a bona fide employee working solely for Consultant, any fee, commission, percentage, brokerage fee, gifts, or any other consideration contingent upon or resulting from the award, making or amending of any determinations with respect to performance of this Contract. Consultant further warrants that gratuities in the form of entertainment, gifts, or otherwise will not be offered or given by the Consultant or any agent or representative of the Consultant, to any director, officer, or employee of the Owner with the view towards distorting the intent of this Contract.

2. For breach or violation of this warranty, Owner shall have the right to annul the Contract without liability and to recover all compensation paid to Consultant prior to such annulment or, at the Owner's discretion, Owner may terminate the Contract without liability, except that prior compensation paid the Consultant shall not be recoverable.

3. Termination or annulment for breach of Warranty against Contingent Fees and Gratuities shall be by written notice, and only after a hearing before the president of the University in which the Consultant shall be invited to participate. Further, the facts presented and findings of such hearing shall be documented by the Owner, and may be issued to the Consultant or for review by any competent Court.

I. WARRANTY AGAINST EMPLOYMENT OF OWNER'S EMPLOYEES

1. Consultant warrants that it has not and shall not engage, during the negotiation or term of the Contract, any personnel who, with Consultant's knowledge, are or have been with two (2) years prior to the effective date of this Contract in the employ of the Owner.

2. In the event Consultant fails to comply with this warranty, Owner may terminate the Contract after a hearing and finding of facts, as described in Item H.3. of this Appendix.

J. WORK WEEK

All work shall be done on the basis of a forty (40) hour work week unless otherwise agreed to by Owner, except in emergency cases where the safety of people or ability to meet the overall project schedule might be endangered without overtime or shift work. All workers performing work in excess of eight (8) hours in any calendar day, or in excess of forty

(40) hours in any work week, shall be compensated at the regular hourly rate unless such worker qualifies under the provisions of the Federal Contract Work Hours Standards Act for a higher compensation rate for overtime.

K. AUDIT OF DIRECT COSTS

1. For the purpose of verifying the actual direct costs incurred by Consultant for any work performed under the Contract or for the purpose of verifying that cost or pricing data submitted in conjunction with the negotiation of this Contract or any Contract Change or other Contract Modification were accurate, complete, and current, Owner, in the person of its authorized agent or its authorized representative, shall be permitted at all reasonable times to confidentially inspect Consultant's books, records, documents, papers, and other supporting data which involve transactions related to this Contract or which will permit adequate evaluation of the direct costs of this work.

2. Any such audit or audits must be completed within six (6) months after completion or termination of the work. If Owner fails to complete this audit within such period, the books, records, documents, papers, and other supporting data as prepared or presented shall be conclusively presumed to be correct.

3. The Owner, in the person of its authorized agent or its authorized representative, may also inspect subcontractors' books, records, and accounts pertaining to the work, provided such subcontract was not based on bids from two or more firms in direct price competition, and further provided that such inspection is conducted prior to six (6) months after expiration or termination of the subcontract.

4. Consultant and subcontractors shall preserve and make available their books, records, documents, papers, and other supporting data until the expiration of the audit periods as herein prescribed.

5. Cost or Price Modification Because of Audit
   If, as a result of audit or otherwise, any price or cost is found to be incorrect or inaccurate, then such price or cost shall be changed to reflect the true price or cost.

L. RESPONSIBILITIES AND LIMITATIONS

Consultant's liabilities for its responsibilities under the terms of this Contract and its Appendices are as follows:

1. Data and Services Furnished by Owner

   Consultant shall have no liability for defects or failure in the work or Project attributable to Consultant's reliance upon or use of data, design criteria, specifications, drawings, or other information furnished by the Owner. This exclusion includes designs by other consultants employed directly by the Owner, despite any reviews, approvals, or acceptance of such design by the Consultant.

2. Data and Services Furnished by Consultant

    After completion of Consultant's services, and in the event of any defect or failure in the work or Project attributable to negligent errors or omissions in the engineering, designs, drawings, specifications, and other data prepared by Consultant or his subcontractors, the Consultant shall perform re-engineering including the preparation of altered designs, drawings, and specifications and will provide construction inspection and monitoring services to the extent necessary without cost to Owner, provided, however, that Consultant's obligations hereunder are limited to those defects and failures which shall be first disclosed within one (1) year after completion of Consultant's services hereunder. Further, such defect or failure disclosure must be through written notice by the Owner to the Consultant within such one-year period.

    The performance of such re-engineering and additional construction inspection shall constitute Consultant's sole obligation and liability to Owner on account of any such defect, failure, error, or omission. Consultant shall not be obligated, on account of any defect or failure, to provide, reinstall, or otherwise alter any equipment, materials, or other portions of the work. Further, Consultant shall not be liable for defects for failures attributable to normal wear and tear.

3. Equipment and Materials

    a. Consultant shall have no liability of any nature for defects in any equipment or materials procured, furnished, or supplied by the Owner, except that if the Consultant specifies, designs, procures, furnishes, or supplies such equipment on behalf of the Owner, Consultant shall, at Owner's request and expense, provide engineering and inspection to correct any defects.

    b. Consultant shall use his best efforts to arrange and assist in obtaining, on behalf of the Owner, all guarantees, warranties, and patent indemnity undertakings to the extent available, but Consultant's liability herein is expressly limited to assisting Owner, at Owner's request and expense.

4. Construction and Workmanship

    a. Consultant shall employ his best efforts to monitor and inspect or, if so authorized, perform the construction or workmanship entering into the Project. Nevertheless, Consultant shall have no liability for any defect in, or failure of any construction or workmanship entering the Project unless such defect or failure is attributable to criminal negligence on the part of Consultant.

b. Consultant shall use his best efforts to arrange for and assist in obtaining, on behalf of Owner, all guarantees, warranties, and indemnity undertakings from construction contractors and other persons engaged by the Owner to perform construction or to render construction services to the Project. The Consultant's responsibility with respect to such guarantees, warranties, and undertakings is to assist and advise Owner, at Owner's request and expense, in obtaining and enforcing of such guarantees, warranties, and undertakings.

5. Consequential Damages

Notwithstanding anything in the Contract or in this Appendix D to the Contract to the contrary, it is agreed that Consultant shall have no liability for loss of revenue, loss of profit, loss of use, or any other indirect or consequential damages.

M. PATENT RIGHTS

Owner shall indemnify and save Consultant harmless from all claims growing out of any patent infringements or claims thereof pertaining to the work based on data, design criteria, or other specifications, drawings, or information furnished by Owner to Consultant.

N. NOTICES

Any notices required to be given herein shall be deemed to have been sufficiently given to either party for all purposes hereof, if mailed by registered mail, postage prepaid, addressed as follows:

1. *To Consultant:*
    CONSULTANT FIRM DESIGN, INC.
    10 Main Street
    Central City, California 99999
        Attention: Mr. I. M. Manager
    AND TO:
    CONSULTANT FIRM DESIGN, INC.
    Project Office
    P. O. Box 5555
    Central City, California 99999
        Attention: Project Manager

2. *To Owner:*
      President,
      CENTRAL CITY UNIVERSITY
      10 Campus Road
      Central City, California 99999
      AND TO:
      Board of Regents
      CENTRAL CITY UNIVERSITY
      40 Administration Building
      University Campus Road
      Central City, California 99999

Appendix E

Insurance

A. Until such time as Consultant's obligations under this Contract shall have been discharged, Consultant shall, at no cost to the Owner, take out, carry, and maintain at least the following coverage and limits of insurance which shall be maintained with insurers and under forms of policies satisfactory to the Owner.[5]

1. Workmen's Compensation and Employer's Liability
    a. State Workmen's Compensation—coverage as required by law.
    b. Federal Longshoremen's and Harbor Workers' Act—coverage as required by law.
    c. Employer's Liability—limits of at least $100,000 per occurrence.

2. Comprehensive Bodily Injury Liability and Property Damage Liability insurance, including coverage for automobiles owned or hired by Consultant, with limits as follows:
    a. Bodily Injury - $500,000 each person
                       -      $1,000,000 each occurrence
    b. Property Damage - $1,000,000 each occurrence (subject to $500 deductible each occurrence).

3. "All Risk" Contractor's Equipment Insurance

B. Owner shall be furnished with a Certificate of Insurance as evidence that the foregoing insurance coverages are in effect. Further, all policies shall contain an endorsement providing that written notice shall be given to the Owner at least thirty (30) days prior to termination, cancellation, or reduction of coverage in the policy.

C. The Bodily Injury and Property Damage Liability policy shall contain the following:

1. A provision or endorsement:
    a. Naming the Owner as additionally insured with respect to liabilities arising out of Consultant's performance of the work under this Contract.
    b. Stating that such insurance is primary insurance as respects the interests of the Owner in the Consultant's work, and that any other insurance maintained by the Owner is excess and not contributing insurance with the insurance required hereunder.

[5]*The clause above in reference to obtaining coverage "at no cost to the Owner" is a negotiable point or option.*

2. A provision or endorsement which eliminates any exclusion or exclusions regarding loss or damage to property caused by explosion or resulting from collapse of buildings or structures or damage to property underground, commonly referred to by insurers as the "XCU" hazards.

3. A "Cross Liability" or "Severability of Interest" clause.

D. The Owner agrees to include in each construction contract for the Project a provision whereby the Contractor is required to take out, carry, and maintain, during the performance of his work under such contract, at no cost to the Owner or Consultant, the following insurance coverages, which shall be maintained with insurers and under forms of policies satisfactory to the Owner:

1. Comprehensive Bodily Injury Liability and Property Damage Liability insurance, including coverage for all operations of Contractor with minimum limits to be mutually determined by the Owner and Consultant, and such insurance to include the Owner and the Consultant as named insureds as respects Contractor's performance of work under such contract. Such insurance shall be primary insurance with respect to Contractor's operations, irrespective of any other insurance maintained by the Owner and/or the Consultant.

2. "All Risk" Contractor's Installation Insurance, covering all work, including temporary facilities, and all materials and equipment to be incorporated in the Project under such contract, including any materials or equipment furnished by the Owner, while at the site of the Project or elsewhere, while in storage and during transit, insuring for a limit of not less than the value of such materials or equipment. Such insurance shall be subject to deductible provisions not to exceed $5,000 each occurrence.

3. The insurance policies for the coverages required in Items D-1 and D-2 above shall name the Contractor, the Owner, and the Consultant as insureds, as their respective interests may appear. Such insurance, and also any Construction Equipment Insurance maintained by the Contractor and applicable to the Project, shall include an insurer's waiver of subrogation rights in favor of the Owner and Consultant, and their officers and employees, respectively.

4. Contractor shall be required to deliver to Consultant, prior to commencement of any field work on the Project, duplicate copies (one copy to be retained by Consultant, and the other to be forwarded to the Owner within five (5) days of receipt) of all policies and certifi-

cates issued by an insurer evidencing compliance with the foregoing requirements. Such policies and certificates shall be in a form and by an insurer satisfactory to both the Owner and the Consultant. Further, such policies and certificates shall provide that written notice shall be given to the Owner and the Consultant at least thirty (30) days prior to the termination, cancellation, or change of coverage in said policies.

E. The Owner shall take out, carry, and maintain insurance which shall be in excess of the required insurance coverages described in the foregoing portions of this Appendix E to the Contract, in amounts and forms mutually satisfactory to the Consultant and Owner, which shall insure against the liabilities and obligations otherwise assumed by the Owner pursuant to this Contract. Alternatively, the Owner may authorize Consultant to provide, at Owner's expense, such additional insurance in mutually satisfactory form and amount. Such insurance shall name Consultant as an insured, and shall contain a "Cross Liability" or "Severability of Interest" clause. Such insurance shall not in any manner limit or qualify the liabilities and obligations otherwise assumed by Owner or Consultant pursuant to this Contract.

In reading over the sample service contract on the preceding pages, you have probably found areas, or perhaps whole sections, to which you object. This is good, for we must remember that this type of contract must be tailored to the needs of the individual project, client, and AECM team; it must be flexible. Yet, it must not be forgotten that this document is the basis for the client-consultant relationship.

In our sample contract, much of the liability for the project was shifted to the owner, Central City University. Perhaps the university may not be willing to accept this liability. If so, it might be desirable to require the consultant to indemnify the owner, and this could be done by modifying this sample contract, adding language such as that shown in the following optional items:

OPTION NO. 1

Indemnification

> Consultant shall assume responsibility and liability for any damage, loss, or injury of any kind or nature whatever to person or property caused by, or resulting from, or in connection with any negligent action, wrongful or improper act, neglect, omission, or failure to act under a duty to act. Consultant shall indemnify and hold harmless the Owner, Owner's representative, and any and all of its officers, agents, representatives, servants, or employees from and against any and all claims, losses, damages, charges, or expense, whether direct or indirect, to which they or any of them may be put or subjected to by reason of any such damage, loss, or injury.

This language, together with other necessary revisions to the sample contract, would materially change the liability situation for the owner and consultant, since the consultant would now be required to indemnify the owner, rather than the owner indemnifying the consultant, as indicated in the sample contract.

A "middle of the road" solution to indemnification can possibly be found in language which sets limits on the consultant's liability and indicates the level or standard for the services to be performed, perhaps somewhat as in Option No. 2, which follows:

OPTION NO. 2

Consultant Responsibilities and Limitations

> A. DESIGN SERVICES
> Consultant shall perform design services, as provided hereunder and as enumerated in Appendix A attached hereto, as an indepen-

dent contractor with that standard of care, skill, and diligence normally provided by a professional architect or professional engineer in the performance of design services.

Consultant's liability for failure to comply with the standards specified above shall be limited to re-performing, at Consultant's expense, any design services which are deficient because of Contractor's failure to meet the above standards in its performance of such services in any portion of the Project, and which deficiencies are made known to Consultant within one year after expiration of the term of this Contract. Notwithstanding the above provisions, it is agreed that this type of work, by its very nature, will require changes and revisions to design and, as such, they shall be performed at Owner's expense unless it can be clearly established that Consultant did not meet specified standards in the initial performance of services. Further, it is understood and agreed by Owner and Consultant that the Consultant shall have no liability to the Owner with respect to Consultant's design services for loss of or damage to Owner's property or additional costs to the Project for any reason (excluding negligence), and the Owner hereby waives any claim for damages therein.

B. PROJECT MANAGEMENT, CONSTRUCTION MANAGEMENT, AND OTHER SERVICES
Consultant shall perform its Project Management, Construction Management, Technical Inspection, Procurement, or other services (excluding design services) as provided hereunder and as enumerated in Appendix A, attached, as an independent contractor with the standards of care, skill, and diligence normally provided by Construction Managers, Technical Inspectors, or other professional or specialists employed in this type work for services similar to those hereunder.

In the event of failure to comply with the standards specified above, Consultant shall re-perform, at the Consultant's expense, any of its own services which are deficient as a result thereof, provided that such deficiencies are made known to Contractor within one year after the expiration of the term of this Contract.

Except as provided for in Section C, Indemnification and Holding Harmless, below, Consultant's total liability to Owner arising out of its Project Management, Construction Management, Technical Inspection, Procurement, or other services (excluding design services), shall be limited to $1,000,000 (not including re-performance costs) for any reason, including Consultant's negligence in performing the services hereunder.

C. INDEMNIFICATION AND HOLDING HARMLESS

Consultant agrees to indemnify and hold harmless the Owner from and against claims for bodily injury, including death, occurring at or adjacent to the Project, brought or asserted by employees or business invitees of Consultant, and Owner agrees to indemnify and hold harmless the Consultant from and against such claims brought or asserted by employees or business invitees of the Owner. "Business invitees" of the Owner shall include construction contractors or subcontractors engaged by the Owner in construction of the Project. "Business invitees" of the Owner shall not include employees of the contractors or subcontractors engaged by the Owner in the construction of the Project.

D. CONSEQUENTIAL DAMAGES

Notwithstanding any of the foregoing, under no circumstances shall Consultant, its officers, directors, partners, employees, or agents be liable to the Owner for any damages for loss of use, loss of revenue, or delays in completion of the Project arising out of, or in connection with its services under this agreement.

These provisions of the foregoing Option No. 2, together with appropriate Insurance provisions, and through proper adaptation to the Project, can be used to satisfy some of the dilemmas of liabilities in this type of contract.

Although this book is not intended to be a legal text or to supplant the advice of legal counsel, we who are in the business of Project Management, Design, Construction Management, or other professional services related to construction must give proper consideration to all contracts, and especially to those professional services contracts.

Even though this sample contract may not fulfill your particular requirements, it does serve us well by providing a model form and also by drawing our attention and interest to some of the intricacies of this important area. It must never be forgotten that the Contract is the principal tool and guide for both Consultant and Owner employers.

Flexibility should be the key word for the AECM team in its form and in its contracting for services. It should also be noted that this flexibility is particularly advantageous to Owner employers.

Chapter 8

Development of Design Services

After selection of the Design Team, negotiation, and final agreement on an appropriate AECM service contract, the next obvious step leads into Planning and Design services.

Unfortunately, a complete description of the complex operations of the design process for even Central City University's new campus project is far beyond the scope of this book, but the objective of this chapter will be to discuss the AECM team during the design phase, to describe certain key activities, to suggest policies, and to describe procedures which are intended to assure successful completion of the design portion of this team's efforts.

For convenience, and as an aid to understanding AECM project work, let us consider that the project team for C. C. University's project is to be organized as illustrated in Figure 8.1 during the design phases of the project.

Note at this point that the positions are indicated for convenience only; reorganizing and regrouping may be necessary several times during the project. Further, these groupings are intended to reflect a single team consulting organization. The dotted lines indicate possible separate divisions of a single firm or groupings of separate firms joined to form the AECMM consulting organization that has been employed by Central City University.

It is not the intent of this text to limit or restrict the freedom of choice and action of either owners, consultants, or contractors in the AECM field. Other organizational or management systems, including different forms or types of AECM organizations, could be successfully employed on this project. The AECMM team organization described in this chapter was set up for illustrative purposes, and because it reflects the personal experiences and preferences of the author.

We observe further that good sense, sound judgment, individual initiative, proper action, and quality professional service are all required of an AECM team and its members, jointly and individually, not only during the design stages, but throughout the entire duration of any project.

TEAM MEMBERS AND THE ORGANIZATION CHART

It is obvious that the size of an AECMM project staff is directly related to the size and complexity of the project at hand. Our example "Design Organization Chart," shown in Figure 8.1, shows what may be considered a normal or typical team structure.

This organization is headed by a Project Manager who may, on occasion, be supported by an assistant. The Assistant to the Project Manager will act for the Project Manager in his absence and will accept the various task assignments the Project Manager delegates. He must, however, act for or in the name of the Project Manager. It is important that the Assistant to the Project Manager operates in this manner. It is obvious that the one-on-one relationship, which would result if the Assistant to the Project Manager breached the organization chart directly below the Project Manager, is not desirable. As a matter of fact, this one-on-one type organization structure would be contrary to good organization theory and very wasteful.

The Project Manager must use common sense, organizational theory, and his good judgment in forming or building an appropriate project team. Since attention and priorities of interests shift throughout a project, the Project Manager must establish himself as a cohesive force and the principal leader within the team.

In the organization for Central City University's project team, the Project Architect, Project Engineer, and Construction Manager lead their respective technical discipline groups. Each of these principal organizational elements must contribute its part to the total project effort, and each of the principal discipline contributions is dependent upon the total cooperation of the others.

During progress of the work, the focus of effort or attention will shift many times among the various participating disciplines, groups, and members of the team. In the early phases of the project, attention may be directed to the Project Engineer, the Project Architect, planners, or members of the specialty consulting staff.

In latter phases of the design, attention will be focused primarily on the Project Architect and/or the Project Engineer (though it may periodically be shifted to other team members), and will gradually shift to the Construction Manager, especially at the end of the design phases or when construction operations are initiated.

Throughout the various design phases and during the construction, it is important that a split in leadership be avoided; team participation and involvement must be fostered, especially by the various principal disciplines. The obligation in this regard inevitably will fall to the Project Manager, who will be required to keep the project unit effectively directed

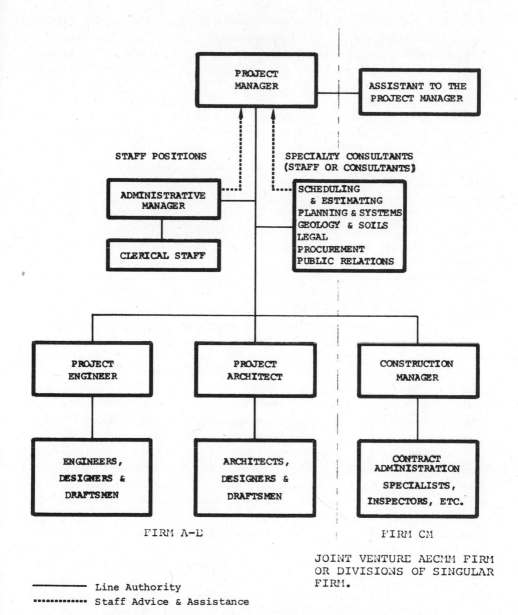

Figure 8.1. Sample Design Organization Chart for AECMM Consultant Firm Design, Inc. staff on Central City University Campus Development.

toward the project goals. This direction will require continued evaluation and establishment of priorities of objectives for himself and for his project team, together with maintenance of an up-to-date awareness of many-faceted project problem areas. The Project Manager must be integrally involved in the team's efforts, which will inevitably include maintaining close liaisons with the team and within the team, as well as with those outside the team. This outside liaison will include the client, construction contractors, and various public officials.

It is essential that the Project Manager recognizes that his responsibilities extend beyond surveillance of the AECM team's work. A Project Manager must be integrally involved with the project and project personnel. He must set objectives and continually establish and define, or redefine, specific performance requirements of the team members or groups. Subsequently, these goal-setting activities must be duplicated by the specific group or discipline leaders, working with and for the Project Manager, in such greater detail as may be necessary within group or subteam organizations under their functional control.

Both separately and jointly, the Project Manager and key members of his project team must insure that: the work performed fits the requirements of the project; the work required is performed properly; costs are considered and adequate cost controls are maintained; and objectives and goals are accomplished in an orderly manner.

To insure the above conditions, key team members, such as the Project Architect, Project Engineer, and Construction Manager, will almost invariably have to establish priorities for themselves and those under their direction. Considerable effort must be spent to direct attention and work on a priority basis, but in so doing substantial dividends are realized.

Numerous tools can be developed to aid in determining what activities must be accomplished first, which are most important, and even which are critical. Even with the diversity of projects to which AECM teams are exposed, a certain continuity and reasonable uniformity may be found through use of effort-directing tools.

The principal effort-directing tool in AECM work is the Consultant-Client contract. Normally, in this principal document, as in our sample contract in Chapter 7, there is a fairly complete and detailed Statement of Scope section. The desirability of distributing the statement of scope of the consultant's services within the AECM team is readily evident, although a complete statement of scope need not be made to all members of the team. The method and the form of the distributed statement is a function of the type, form, and personnel in the AECM team and their need for this information. In certain teams, it may be desirable to convey the scope information through distribution of the entire service contract or entire parts of it. In other cases, it may be desirable to restate the scope of services in a new form or forms which are designed to convey the necessary understanding of the services that either the team or the individual team member is expected to perform.

The objective of a scope statement is to define the consulting services in the required detail and to establish limits on the work effort. Reasons for distributing a scope statement within the AECM team include: development of a spirit of involvement; elimination of rework which can result from misunderstandings regarding the scope or performance requirement; and initiation of control by drawing attention to scope and scope changes.

Once the scope is defined, distributed, and understood, it is essential that the services indicated be adhered to as closely as possible. It is of utmost importance that the appropriate AECM team members and the client be promptly advised when scope changes have been requested or have evolved.

Scope changes are most likely to occur in the early stages of a project when major design decisions are being made. Members of the AECM team should, therefore, be particularly alert to this possibilty during this period. Additionally, it is imperative that an orderly and complete record of changes and change requests be kept and that the appropriate team members and the client be kept appraised unequivocally of such changes. Such appraisals should include a complete presentation of what these scope changes are, together with estimates of the cost, time, and other implications of such changes.

When scope changes are made, especially those which expand the work to be performed under the AECM Consultant Services Contract, it is highly desirable to set up cost accounting procedures that will allow the identification of the additional costs and to make such contract changes, modifications, or supplemental agreements as are appropriate.

The initial or set-up work for design has much to do with the successful completion of all phases of AECM work. For our purposes, let us consider the design effort, consisting of four major phases, as follows: Planning—Conceptual and Schematic, Design Development, Contract Documents, and Services for Construction.

If we do this, then we are able to describe, in general, what work is included in each phase and what work must be complete at the end of each phase. Many times, some specific performance requirements of these phases, together with other performance requirements, are found in the AECM service contract. If this is the case, it is essential that these contract requirements be satisfied. Often, these contractual requirements can be integrated into a flexible outline of performance requirements by design phase developed to give direction, to prevent misunderstandings, and to serve as a performance checklist.

Since AECM work can include such a diverse field of endeavor, it is impractical to attempt here to develop checklists for every type of project. Insight into this work should be developed, however, by careful study of the two following checklists. These are AECM Building Type Contract Design Phase Performance and Accomplishment Checklists for (I) Planning—Conceptual and Schematic Phase and (II) Design Development Phase. Activities and checks for the other two major phases—the Contract Documents Phase and the Services for Construction Phase—will be discussed in detail in later chapters of the text.

I.   CHECKLIST FOR PLANNING, CONCEPTUAL AND SCHEMATIC

    A. GENERAL
        1. Obtain registered surveys of property. This should include establishment of property lines and corners; the general contours, easements, and locations of above and below ground utilities; and easements.
        2. Obtain photographs of the site and nearby properties.
        3. Obtain copies of the applicable codes and governing insurance association requirements.
        4. Establish the design criteria for *all* disciplines.
        5. Order subsurface investigations, soil borings, etc.
        6. During this phase, give definite consideration to costs and materials. Consider, especially, the building space, the exterior wall treatment, and the mechanical system. Identify special material and area requirements.
        7. Attempt to select the best possible structural, mechanical, and electrical systems to support the architectural concept.
        8. Indicate future expansion and give proper consideration to this.
        9. Identify the need for using the services of consultants from outside the AECM team.
       10. Establish the use of single line drawings.
    B. DRAWINGS
        1. *Site plans*—should include:
           a. Location of proposed building(s).
           b. Outline location of existing structures, utilities, and special features, together with an indication of whether they are to be removed or to remain.
           c. Location of utility areas.
           d. Contour map(s) of site.
           e. Location of property lines.
           f. Location and names of bordering streets, highways, or railroads and connections or access to these.

     g. Block layouts of parking lots, roads, and other major land uses, such as athletic areas, foot paths, drainage fields, etc.

     h. Limits of contract(s).

2. *Architectural plans*—should include:
   a. Floor plans with overall outside-to-outside building dimensions.
   b. Approximate bay spaces.
   c. Rooms and their identities.
   d. All mechanical and electrical equipment rooms, shafts, and related spaces or areas.
   e. Locations of elevators, escalators, dumbwaiters, and stairwells.
   f. A penthouse plan and roof plan showing major projections above the roof.

3. *Architectural elevations*—should indicate:
   a. Massing of the building or buildings.
   b. First floor elevation and, if possible, a typical floor elevation.

4. *Architectural sections*—should indicate:
   a. Any unique features or conditions.
   b. Approximate floor heights and ceiling heights.

C. TECHNICAL DESCRIPTION, DESIGN CRITERIA, AND OUTLINE SPECIFICATIONS—should be prepared.

The Outline Specifications should be prepared on the basis of a uniform system similar to the specifications to be employed as Contract Documents, and they should include:

1. Statements of all architectural and engineering design criteria (architectural, mechanical, electrical, civil, structural, etc.).
2. Description of the structural system, including a description of construction materials and probable construction methods.
3. Description and/or outline specifications for interior and exterior architectural materials.
4. Outline specifications or general descriptions of the following types of work or systems.
   a. Civil.
   b. Heating, ventilating, and air-conditioning.
   c. Plumbing.
   d. Fire protection.
   e. Refrigeration.
   f. Power.
   g. Lighting.
   h. Communication systems.

i. Fire and/or other alarm systems.
j. Special electrical systems.
k. Special utility systems.
l. Transportation systems (elevators, escalators, etc.).

D. COST ESTIMATES, BUDGETS, OR COST APPROXIMATIONS

These should include all building components or features, and allowances must be made for:

1. Growth and design development contingencies.
2. Cost escalation.
3. Construction and design contingencies.

E. APPROVAL, RELEASE, AND LETTER OF TRANSMITTAL

1. Cognizant members of the design effort, including the project manager, should date, sign, and approve all work to be released to the client for approval.
2. Release to the client should be handled through a letter of transmittal which should:
    a. Itemize the documents being transmitted.
    b. Request client's written approval, pointing out the date approval is expected or required so that the project schedule can be maintained.
    c. Client approval should include approval of:
        (1) Drawings.
        (2) Technical description, design criteria, and outline specifications.
        (3) Cost estimates, budgets, or cost approximations.
        (4) Building areas.

II. CHECKLIST FOR DESIGN DEVELOPMENT

A. GENERAL

1. Before starting the Design Development phase, the following items are required:
    a. *Written* approval of the Planning, Conceptual and Schematic phase (or phases) of design. (Note that, for our purposes, we have lumped these potentially distinct phases into one initial phase in this text.)
    b. A complete registered survey of the property.
    c. A complete soils investigation of the property, including soil borings for all footing locations.

2. Design drafting or drawing for this phase should:
   a. Be started on standard size tracings which are intended to become the tracings for the working drawings at the appropriate time.
   b. Avoid overdrawing, excessive hatching, excessive shaping, etc.
   c. Be coordinated so that all plan views are oriented in the same direction or on the same grid system and lines.
3. During the initial part of this phase:
   a. A list of drawings and an index of specifications should have already been, or should now be prepared.
   b. Consultants' drawings should be started. It is often desirable to require substantial completion of these drawings early so that proper coordination can be obtained between the various design disciplines.
   c. A thorough and documented check of applicable codes, regulations, and laws should be made. Liaison with local governmental agencies, including fire marshals and building departments, should be established. All design reviews or interpretations, including those made by government officials, should become part of the project's written records.
4. Throughout this phase, particular attention must be given to costs:
   a. Cost trends should be identified and appropriate controls established.
   b. Major changes or revisions to what was provided for in the prior design phase(s) should be reviewed.
   c. New features should be identified.

B. DRAWINGS

By discipline, the drawings should include, reflect, or indicate the following:

1. Civil
   a. Vicinity map, if appropriate.
   b. Limits of contract.
   c. Storage and utility areas for use of contractor(s).
   d. Complete registered survey of property. Such print(s) should identify the firm that executed the survey and also should bear the *stamp* and *signature* of the responsible engineer or surveyor.
   e. Established bench marks.

    f. Property lines and easements.

    g. Existing and new contours.

    h. Finish grades near buildings.

    i. Layouts of *all* utilities (plus possible access to these facilities for construction).

    j. Railroad, road, or parking lot layouts, if applicable.

    k. Outlines of existing structures.

    l. Existing trees.

    m. Soil borings data.

    n. Standard and key details.

2. Architectural

    a. Floor plans, including:

       (1) Exterior dimensions.

       (2) Column line or grid identification system.

       (3) Bay spacing and room dimensioning.

       (4) Room layouts and room titles.

       (5) Door swings.

       (6) Fixed equipment layouts.

       (7) Furniture layouts, if applicable.

       (8) Indications of items which are:

          (a) Furnished by owner to be installed by contractor.

          (b) Furnished by contractor.

          (c) Neither to be furnished nor installed by contractor, using the indication "Not in Contract" (NIC). Note that, if such NIC items are to be installed by others during the life of a contract, the contractor should be so advised.

    b. Roof plans, including:

       (1) All major projections above the roof, such as a penthouse, mechanical enclosure, or cooling tower.

       (2) Roof slopes and roof drains.

       (3) Access to roofs.

       (4) Expansion joints.

       (5) Hoisting and/or window washing devices, if applicable.

    c. Exterior elevations, indicating:

       (1) Building exterior appearance including windows, window walls, wall elements, or ornamentation.

       (2) Wall materials.

       (3) Control and expansion joints.

       (4) Floor-to-floor heights.

       (5) All projections above the roof.

   d. Building sections, indicating:
      (1) Floor-to-floor heights.
      (2) Ceiling heights.
      (3) Structural system(s).
      (4) Typical wall sections, both exterior and interior, including expansion joints and other critical details.
      (5) Section studies at critical points, such as transitions from vertical to horizontal runs for conduits or duct out of shafts.
   e. Reflected ceiling plans.
   f. Interior elevations, especially for lobbies and major walls.
   g. Preliminary schedules, including:
      (1) Door schedule.
      (2) Room finish schedule.

3. Structural
   a. Foundation plans indicating type of foundation and extent of basement.
   b. Floor plans:
      (1) Indicating structural framing system with typical sizes of beams and columns.
      (2) Where critical areas exist, detailed studies of beam and column sizes should be made.

4. Mechanical
   a. Plumbing drawings, including:
      (1) Riser diagrams.
      (2) Toilet room piping layouts.
      (3) Special systems of features.
   b. Heating, ventilating, and air conditioning drawings, including:
      (1) Single line diagrams for each typical floor.
      (2) Major duct runs indicating sizes.
      (3) Block layouts of mechanical rooms and HVAC equipment.
      (4) Location of diffusers.
   c. Fire protection drawings, including:
      (1) Location of fire hose connections.
      (2) Location of fire hose cabinets.
      (3) Location (or typical location) of sprinkler heads.

5. Electrical
   a. An overall single line diagram.
   b. Lighting plans.
   c. Layouts of all electrical areas, including locations of light-

          ing panels, telephone panels, power panels, substations, etc.

    d. Typical underfloor duct layouts.

    e. Transportation drawings showing elevators, escalators, conveyors, people movers, etc. (Note: Transportation drawings, plans, and details require special multidiscipline coordination, especially between architectural, structural, and electrical drawings.)

C. DOCUMENTATION

Although not necessarily for extensive distribution during the design development phase, extensive documentation is required by the disciplines involved. Effort must be spent on such things as design criteria tables and computation summary booklets. Further, all engineering design computations should be prepared in an ordered and systematic format.

D. MANUFACTURERS' BROCHURES

A thorough review of manufacturers' brochures and other literature should be made for:

1. All special equipment, such as computer decking, printing or reproduction machinery, laboratory equipment, etc.
2. Finished hardware.
3. Ceiling systems.
4. Light fixtures.
5. All exposed electrical equipment.
6. All exposed mechanical equipment, such as drinking fountains, thermostats, etc.
7. Plumbing fixtures and toilet room accessories.

E. RENDERINGS AND/OR MODELS

These should be developed if required, or if the cost of such work is justifiable.

F. OUTLINE SPECIFICATIONS

These should be developed, using a uniform system; particular attention to detail is justified in this effort.

G. COST ESTIMATING AND TRENDING

A continuous effort is required in this area through the design development stage. As completion of this stage of design approaches, a reasonably comprehensive estimate of costs can be made from a compilation of available information. Such an estimate should be detailed and should indicate the quantities and quality of work on which it is based. It should also include:

1. A cost escalation factor.
2. Contingencies for both design changes prior to construction and for changed and extra work during construction.

H. APPROVALS, RELEASES, AND TRANSMITTAL OF COM-PLETED WORK

1. The cognizant designers and managers should be required to provide *signed approval* of completed design development work.
2. Those documents which are to be released by the AECM team for client review should be identified, dated, and approved by the project manager or responsible designer.
3. All documents sent to the client must be covered by a letter of transmittal which should:
    a. Itemize all documents being transmitted.
    b. Indicate schedule status, review time, and the date approval is requested and/or required.
    c. Request *written approval* of:
        (1) Design development drawings.
        (2) Outline specifications.
        (3) Cost estimate(s).
        (4) Renderings or model, if applicable.
        (5) Any special enclosures, such as manufacturers' or other brochures, current schedules, status reports, etc.

With completion of Design Development, the emphasis is gradually shifted from the purer design disciplines to the practical world, which includes the areas of contract administration, law, project management, and construction. This is not to say that design is complete for indeed it may not be. Design effort will be required in the Contract Document phase and possibly even in the Services for Construction phase.

Chapter 9

Contract & Project Services

For our purposes, and for continuity, consider Contract and Project Services to include the phases of total AECM called: Contract Document Phase, and Project Services Phase.

We, of course, realize that there must be continuity in services, and that establishment of phases or milestones throughout the project facilitates control. It is during these two phases of the project that unprecedented attention to detail is required.

Every drawing should be checked for completeness and correlation with every other drawing. Every word of the specifications should be reviewed to be sure the specifications are exact, complete, and free of duplications or grey areas. It is well to remember that the completed contract documents will serve as the primary tools in relation between construction management and contractor or subcontractor.

A description of the Contract Document phase follows, along with a sample checklist for this phase. It is hoped this checklist will provide continuity of effort by tying together this phase with the two preceding phases. A finalizing narrative on the Project Services Phase is introduced for development in Chapter 10.

Additionally, it is noted that since the checklists for the Planning— Conceptual and Schematic Phase, Design Development Phase, and Contract Document Phase are presented in a complementary format, the reader may well want to mark the pages, assemble them in order, or otherwise compile them for ease of reference.

CHECKLIST FOR CONTRACT DOCUMENT PHASE

   A. *GENERAL REQUIREMENTS*
      1. Written approval of the Design Development documents must

be obtained from the client before starting the Contract Document Phase.

2. Utilization of drawings prepared in earlier phases must be made to the fullest extent possible.

3. All specifications should be a continuation of the outline specifications prepared in the Design Development Phase. Construction management personnel should be encouraged to contribute to and proof all specifications.

4. A listing of all anticipated drawings should be prepared at the start of this phase.

5. All drawings should be prepared in a standard format, style, and sheet size, all of which are intended to reflect a consistency of effort.

6. The layout of the building(s) on site plans, floor plans, and partial plans should be coordinated to insure that the buildings are oriented in the same direction throughout the set of drawings.

7. Interdisciplinary correlation checks of drawings and specifications should be executed as necessary prior to completion. At least two complete checks should be made.

8. Quality of work should not be sacrificed to the pressures of time or performance, especially near completion of this phase.

B. CIVIL DRAWINGS AND SPECIFICATIONS

  1. General Requirements

    a. The civil drawings should show or identify:

      (1) The location of the site.

      (2) The overall and detailed project concept.

      (3) Physical dimensions of the property.

      (4) Existing and proposed facilities, including:

        (a) All utility services, underground and overhead.

        (b) All paved or surfaced areas.

        (c) All landscaped areas.

        (d) Railroads.

        (e) Fences.

        (f) All buildings.

        (g) Limits of the construction.

      (5) All area plans should indicate North.

      (6) The scale used on the drawings should be indicated.

      (7) Legend(s) for identification of symbols should be provided.

      (8) All soil borings should be located, reference num-

bered, and a tabulation of data should be provided, either on the plans or in the specifications.

    b. Civil specifications must be correlated with other contract specifications. Use of master specifications and/or coordination of the civil specifications with local, state, federal, or other standard heavy or highway specifications should be encouraged.

2. Site Plans

    a. General Site Plan should indicate the overall scope of the project including, if applicable, a vicinity map identifying the location of the project relative to the area.

    b. Detailed Site Plans should indicate in detail the features of existing work together with all physical features of the new site work, including:

       (1) A Key Plan identifying the detailed plans.

       (2) Survey Plans, which should:

          (a) Provide and outline metes and bounds description of the property on which the building(s) will be constructed.

          (b) Locate and provide elevations of established monuments or benchmarks.

          (c) Establish a coordinate system and method of locating and identifying buildings, building features, utilities, and miscellaneous site features.

          (d) Locate soils borings.

          (e) Indicate rights-of-way and/or easements, if applicable.

       (3) Grading plans shall include both rough and fine grading and other information such as:

          (a) Existing grades in areas to be graded and in areas not affected by the project.

          (b) Contour mappings of and/or sections through areas to be graded.

          (c) Finish grade elevations of all surfaces.

          (d) Location of all existing paved areas, utilities, and miscellaneous site features or improvements.

          (e) Identification of all new and existing paving, including finish floor elevations and type of new paving, if applicable.

          (f) Location, type, size, elevation, and other details for all retaining walls, fences, guard houses, artificial lakes, pools, water towers, boiler houses, and electrical substations.

(g) Identification of railroad facilities.

(h) Correlation of detail, especially with landscaping and utility drawings.

(4) Utility Plans shall provide information on the location, type, size, slope, and depth of all utilities, existing and proposed, including:

(a) Separation of utilities by class, if appropriate.

(b) Underground and aerial power and telephone services, including duct banks, manholes, poles, light standards, and substations.

(c) Location and size of gas service piping, including valves, meters, and meter vaults.

(d) Fuel or gas piping systems, including pads and tank form details, pressure information, and other pertinent data.

(e) Storm, sanitary, and industrial sewer systems, including inlets, manholes, headwalls, catch basins, and lift stations.

(f) Water and fire protection piping systems, including inlets, pipes, hydrants, valves, valve boxes, meters, meter pits, post indicator valves, hose boxes, and live connections to existing systems.

(g) Surface drainage systems, including valley gutters and drainage ditches.

(5) Civil and Site Details are basically to provide:

(a) Paving and surfacing details, including details which indicate type and thickness of concrete or bituminous paving and aggregates, surface finish, reinforcing, reinforcing dowels, joint details, joint layouts, curb details, sidewalk details, outdoor steps and railing details, railroad crossing details, and other paving-related details.

(b) Utility details, including details for pipe trenching, bedding and backfill, inlets, catch basins, cleanouts, manholes, meters, meter boxes, distribution and valve boxes and pits, lift stations, septic tanks, leaching fields, headwalls, rip-rap layouts and other erosion control methods, fire protection and water piping anchor and thrust block details, pipe guards, cathodic protection systems, control unit pads and details, and other utility-related details.

(c) Landscaping and lawn sprinkler details, including details of sodded, seeded, and planted areas, and installation details for sprinkler systems.

(d) Miscellaneous site details, including details of fencing, signing, large-scale layouts of site areas, and other miscellaneous details.

C. ARCHITECTURAL DRAWINGS AND SPECIFICATIONS

1. General Requirements

   a. Coordinate with all other design disciplines, especially structural, electrical, and mechanical.

   b. Incorporate changes and make corrections to drawings as quickly as possible as work progresses, and communicate such changes and corrections, preferably by furnishing prints to others.

   c. Schedule completion of drawings and establish cutoff date(s) by which changes must be agreed to.

   d. Avoid overdrawing use schedules; guide details and incorporate these into specifications where possible, or in construction detail book.

2. *Floor Plans* should:

   a. Indicate column center lines or work points, perimeter walls and lines, and subdivisions of the space within the perimeter walls.

   b. Include a key plan, if appropriate, and major building section references.

   c. Include a North arrow.

   d. Show control dimensions of all elements.

   e. Use a common reference for all space dimensioning as from work points, column lines, or perimeter building lines.

   f. Indicate finish floor elevations and floor depressions.

   g. Show changes in floor finishes.

   h. Show walls and partitions drawn to thickness and identified as to type.

   i. Indicate height above ceiling if required for any reason.

   j. Indicate doors, door swing, and door number, and reference doors to door, hardware, and finish schedules.

   k. Show stairs, toilet rooms, elevator enclosures, hatchways, etc., by a larger scale plan if need be.

3. *Roof Plans* should indicate perimeter, roof curbs and parapets, penthouses, special cleaning equipment, openings, etc. Check to insure inclusion of:
   a. A key plan.
   b. A North arrow or orientation reference(s).
   c. Detail references.
   d. Building section references.
   e. Roof slopes and saddles.
   f. Changes in roof materials.
   g. Roof sumps, drains, scuppers, etc.
   h. Grid lines for common reference.
   i. Major items mounted on roof.
   j. Hems penetrating roof.
4. *Building Sections* should indicate the general construction, including footings, foundation walls, floor heights, floor elevations, ceiling heights, plenum space, and floor construction.

   Sections should be cut through various elements as required to insure clarity, such as stairs, elevators, etc. The inclusion of interior room elevations and room names, however, should be avoided.
5. *Elevations* should show all exterior surfaces of the building, including:
   a. Entrances and materials.
   b. Grades at building lines.
   c. Floor and roof lines and elevations of same.
   d. All elements of the facade, labeled and indicated.
   e. Identity of types of sash, glass, brick, architectural concrete, precast concrete, curtain walls, louvers, doors, etc.
   f. Detail references.
   g. Wall expansion joints, control joints, reveal strips, and construction joints, if appropriate.
   h. All major projections above the roof.
6. *Exterior Details* should include detailed vertical and horizontal sections through elements in the exterior walls. Manufacturers' stock items can be indicated with profiles. All materials should be labeled and shown in relationship to other materials. Use of specification is desirable. Details should be provided on:
   a. Concrete cast-in-place, precast, architectural, and structural.
   b. Curtain walls.
   c. Sash.
   d. Glass and glazing.

    e. Doors and frames.

    f. Masonry.

    g. Fastenings.

    h. Caulking.

    i. Waterproofing and dampproofing.

    j. Control, expansion, siesmic, and construction joints.

    k. Roofing, roof openings, and flashing.

    l. Other details for plazas, approaches, and the building exterior.

7. Interior Details, Elevations, and Part Plans.

Interior details include large scale vertical and horizontal sections through elements in the interior of the building. Where necessary, interior elevations and large scale part plans are used to accompany and complement interior details, especially in complicated areas of the building. Details should be checked to insure they are provided on:

    a. Floors.

    b. Interior walls and wall framing.

    c. Elevators, elevator hoistways, and building core.

    d. Stairs.

    e. Ceilings.

    f. Millwork.

    g. Doors and frames referenced to or linked to the door schedule.

    h. Toilet rooms and accessories.

8. *Finish Schedules* can be incorporated into specifications, separately bound in a book of details, or included in plans. The purpose of the Finish Schedules is to indicate all floor, base, wall, and ceiling materials as they occur in each room. The architectural drawings show a name and number for each room or space. The Finish Schedule can be keyed to these room names or numbers, or an alternative system which the author prefers may be employed to key ceiling, wall, base, and floor materials to an alpha-numeric marking by room. Room name, number, and finishes can be very successfully indicated on plan views as noted in Figure 9.1.

Schedules and codes should be thoroughly reviewed to insure:

    a. All materials and finishes are properly designated.

    b. Finishes as scheduled are correlated with details and specifications.

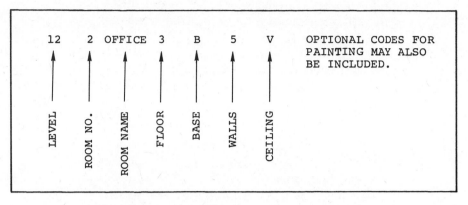

Figure 9.1. Alternative system for keying Finish Schedule, using an alpha-numeric marking by room.

     c. Schedules are complete enough so that construction can be facilitated; materials and finishes must be easily determinable.
  9. *Door and Hardware Schedules* can also be included advantageously in books of details or specifications. Such schedules should:
     a. Indicate materials of doors and frames.
     b. Specify door types.
     c. Show hardware.
     d. Indicate door number, type, and size.
     e. Provide typical and special details.
     f. Indicate fire labels.
 10. Equipment, Casework, and Miscellaneous
     a. Equipment should be indicated complete with layouts, elevations, details, and schedules. These should be checked for thoroughness and completeness.
     b. Casework should be handled the same way as equipment.
     c. Items to be provided under separate contract, furnished by owner, or added later should be so indicated.
 11. Specifications
    Architectural Specifications can often prove to be sources of confusion and difficulty in contract administration and, therefore, deserve special checking and coordinating effort.

    Utilization of Master Specifications, available from specification associations, computer services, and professional associations, should be encouraged, provided they are thoroughly reviewed and tied to the project requirements.

All-encompassing phrases and language should be avoided, as should both cutting and pasting or copying from a vendor's or manufacturer's sample specifications. If such copying must be done, "equal" or "approved equal" products should be provided for.

Clarity and completeness are required in specifications, and much effort may be required to insure these qualities.

D. STRUCTURAL DESIGN AND SPECIFICATIONS
1. General Requirements
   Structural Design must be completely dimensioned and correlated with Architectural, Mechanical, and Electrical Design. Structural drawings must show clearly the work, and they must employ sufficient details and contain notes required to fully explain the design and construction features.
2. Foundation Plans shall show the foundation, footings, foundation walls, area walls, bank support, underpinning, and subdrainage. Check drawings to insure:
   a. Foundation materials and foundation systems, such as piling and caisson work, are properly indicated. It is desirable to show correlation with or reference to soil borings or civil drawings to indicate soil conditions or profiles.
   b. The location of adjacent buildings and their foundations is shown, especially if underpinning or bank support is required.
   c. For pile foundations, the following information is given:
      (1) Elevation of top of footings or caps.
      (2) Elevation of pile cutoffs.
      (3) Designed capacity of piles.
      (4) Estimated length, tied to specifications and/or pay quantity language.
      (5) Piling details such as dowels, moment connections, splicers, welding details, and tip details.
   d. Caissons or foundation piers are treated like piles foundations.
   e. All foundation concrete work is shown and both typical and special details provided, especially if these are expected to have an effect on construction sequence. The dimensions of all concrete are to be provided, together with the location and spacing of reinforcing. Both drawings and schedules can be used for this purpose.

    f. Foundation plans are correlated with mechanical and electrical plans. All pits and trenches shall be shown and dimensioned on the foundation plan(s). These dimensions must be carefully correlated with those given on other plans.

    g. Openings or sleeves for all pipe and ducts passing through the footings, grade beams, and below grade interior or exterior walls are shown.

    h. Miscellaneous details such as dampproofing and waterproofing can often be included conveniently on foundation plans.

3. Framing Plans shall show building line(s), concrete columns, walls, pilasters, girders, beams, slabs, or structural floors, including design loads for each floor or frame. In addition, framing plans should contain, or be checked to insure that:

    a. The structural system consisting of columns, beams, girders, slabs, etc., is clearly indicated. All beam and column sizes, including those for reinforcing steel, prestressing elements, stress rods, etc., shall be listed in the schedules.

    b. Structural steel beams are indicated in accordance with the abbreviations listed in the current AISC Manual. Connections are to be properly indicated and reference made to appropriate details. Welds shall be properly indicated to distinguish between shop and field welds.

    c. Column schedules show total loads on each column. Prestressed beam schedules should indicate required stressing tensions, geometry, anchor plates, and anchor locations.

    d. All structural details are compared to and correlated with current applicable codes.

    e. Stair wells, elevator shafts, access hatches, and related elements are shown and dimensioned on framing plans. These dimensions shall be carefully correlated with those given on Architectural, Mechanical, Electrical, and other "discipline" drawings.

4. Structural Details

Sufficient details to show clearly the design and construction features of the project shall be provided. Details should be organized, grouped, and indicated in a manner that will reduce confusion.

Consideration of construction methods and techniques should be given in preparation of details, especially in relation to connections, reinforcing, prestressing, and precast concrete work.

5. Specifications

Maximum use of standard or master specifications should be encouraged. Nevertheless, if required, special specifications that fit structural requirements such as those for form removal, shoring, reshoring safety, joints, and sequencing should be considered and, if not appropriately noted on plans, integrated into the specifications.

E. MECHANICAL DESIGN AND SPECIFICATIONS

1. General Requirements

The mechanical drawings shall clearly define the extent of the work and contain sufficient details, sections, diagrams, notes, and schedules to fully explain the design and construction features.

Designs must be correlated with those of other disciplines. Categorization of mechanical work into work packages by trades such as HVAC, plumbing (which is preferably subdivided), fire protection work, and landscape sprinkling, in accordance with local craft practice, is desirable.

2. Heating, Ventilating, and Air-Conditioning

a. The HVAC drawings shall be correlated with all other drawings and shall show:

(1) Equipment rooms.

(2) Ductwork.

(3) Piping.

(4) Diagrams which can include flow, piping, and systems diagrams.

(5) Control systems.

(6) Refrigeration systems.

(7) Details, dimensions, and schedules.

b. Particular attention should be paid to "trade" interfaces and to providing detailed information on items like the following:

(1) Ductwork shall be shown complete and to scale with dimensions given on all critical duct locations so that shop building can be used as extensively as possible. All volume and fire dampers a.id all turning vanes shall be shown, including appropriate details.

(2) Insulation shall be shown on drawings unless specifications and/or separate detailed booklets can indicate the insulation clearly.

(3) All equipment and piping should be shown to scale based on selected equipment. In the equipment,

boiler, incinerator, and other rooms, clearances for construction and for access and maintenance shall be shown.

(4) Bases, special pads, and vibration isolation for equipment shall be shown or noted.

(5) Basic mechanical details such as those for pressure reducing stations, meter installations, tanks, pumps, pipe anchors, and pipe expansion shall be indicated.

(6) Underground heat distribution pipe shall be drawn with pipe sizes indicated.

(7) Provisions for anchorage of heat distribution system elements should be detailed.

(8) Flow diagrams should be provided in which equipment is identified and pipe sizes are shown.

(9) Automatic temperature control diagrams and notes should be provided which are complete. Notes shall describe the operation of the system and the desired temperature control shall be indicated on diagrams provided for each type of system in the project.

(10) Air-conditioning units, fans, coils, pumps, valves, etc., shall be identified and scheduled, and requirements shall be listed.

3. Plumbing

The plumbing drawings shall be correlated with all other "trade" drawings and shall show the equipment and piping for all systems other than HVAC and fire protection. Detailed information such as the following should be included.

a. Riser diagrams for water, waste, vent, and other piping systems shall be provided.

b. Sections and elevations should be provided where necessary to clarify design. Critical areas should be detailed at a scale large enough to define and illustrate.

c. All sewer systems (storm, sanitary, combined, or special) shall be shown complete, including pipe sizes, direction of flow, and invert elevations.

d. All plumbing fixtures are to be identified and pipe sizes shown in toilet and service rooms.

e. Clearance dimensions, necessary structural supports, piping connections, and control devices should be shown in details.

f. Special equipment and systems for chemicals, laboratories, hospitals, etc., shall be shown and detailed.

Special effort is often justifiable to clarify installation requirements such as special methods, required performance, or the establishment of "jurisdiction" or responsibility.

   g. Pumps and water heaters should be identified and requirements listed in schedule form.

4. Fire Protection

The fire protection drawings should clearly show the extent and type of sprinkler system to be provided. Fire protection drawings are to be correlated with all other "trade" drawings. *Special effort should be expended to insure that this design is compatible with the fire codes and requirements of the local fire marshal and/or fire department.* Provide detailed information on:

   a. Fire department connections, including location and identification. Insure compatibility with local fire department connections.

   b. Sprinkler systems, including pipe sizing, sprinkler head indication, and review for compliance with governing codes.

   c. Fire pumps and/or storage reservoirs. Location and connecting piping should be shown.

   d. Fire hose and valve cabinets.

   e. Special fire protection equipment or systems.

5. Specifications

Maximum use of Master Specifications should be encouraged, but a thorough review and correlation of specifications with details and other drawings is also to be made.

F. ELECTRICAL DESIGN AND SPECIFICATIONS

1. General Requirements

Electrical design must be correlated with all other "trade" or discipline drawings. They must be designed, reviewed, and coordinated with service power facilities and the authorities of the local power utilities.

Electrical plans are conveniently prepared in the following groupings:

   a. Site and Structure Duct Plans.

   b. Power and Distribution Plans.

   c. Lighting and Distribution Plans.

   d. Auxiliary Systems.

   e. Diagrams, Schedules, and Details.

2. Site and Structure Duct Plans

   a. Drawings shall locate all site-related electrical work, both aerial and underground. All duct work shall also be shown, such as underfloor duct, trench duct, header duct, etc.

   b. Drawings shall indicate the use of each duct run.

   c. Details of crossing ducts, connections to headers or trench duct, connections to panels, and special arrangements shall be provided.

   d. Drawings shall locate all outlets and indicate use, such as power, telephone, etc.

3. Power and Distribution Plans

   a. Correlate plans with local utility companies to insure compatibility.

   b. Provide one-line diagrams of all major power equipment, such as:

     (1) Primary switchgear.

     (2) Switchboards.

     (3) Tie breakers.

     (4) Unit substations.

     (5) Motor control centers.

     These diagrams should indicate number and size of conduit and conductors, voltage, amperage rating, and frame sizes of breakers and fusible switches. One-line diagrams should also indicate trip amperage ratings of fuses to be installed. Transformers should also be indicated either on the diagram or by referring to them in an included or separate schedule. Transformer information should include designations, KVA rating, primary and secondary mode, type, and mounting information.

   c. Power plans shall be correlated with all other "craft" drawings and shall locate such items as primary switchgear, unit substations, power panels, light panels, switchboards, motor control centers, motors, receptacles, and transformers.

   d. Elevations of equipment shall be indicated on the drawings.

   e. Power system riser diagrams shall include the complete system from the incoming service line to the distribution panels. Diagrams shall indicate conduit size, number of wires installed, and wire sizes.

   f. Drawings shall indicate power and lighting panels with reference to schedules.

4. Lighting and Distribution Plans

   a. Lighting system design shall include lighting schedules designating type; alphabetic, alpha-numeric, or numerical designations; lighting schedule(s) indicating wattage of fixtures, type of lamps, operating voltages, and types of mountings, including all special mounting arrangements.

    b. Lighting plans shall locate all lighting fixtures, indoor and outdoor, exit and emergency lighting, and all special effect lighting.

    c. Lighting drawings shall also indicate all switching arrangements.

  5. Auxiliary Electrical Systems

    a. Auxiliary systems shall be shown either separately or integrated with other electrical design.

    b. Drawings shall locate all outlets of the various auxiliary systems such as telephone, intercommunication, fire alarm, burglar alarm, television, music, and other systems resulting from building use requirements.

    c. All auxiliary systems shall have, or be included in a riser diagram(s). Drawings shall show conduit sizes and number and size of wire installed.

    d. Drawings shall also indicate all special mounting details and other details necessary for accurate interpretation.

  6. Specifications

    Use modified Master or Standard Specifications if possible; be detailed and exact.

G. TRANSPORTATION DESIGN AND SPECIFICATIONS

Transportation design shall cover all elevators, materials handling equipment, escalators, people movers, automatic materials moving equipment, etc.

  1. General Requirements

    Plans and details shall be included in Architectural Drawings, yet they must be correlated with all other drawings; especially, the Structural and Electrical Drawings shall show information such as:

    a. Plan(s) of elevators, elevator pits if any, hoistway(s) showing floors served, floors in blind portion of hoistway, secondary levels, and machine rooms with access thereto. Plans shall include clearance and run-by dimensions, platform size, counterweight space, and door space.

    b. Sections through elevator hoistways, pits, secondary levels, and machine rooms.

    c. Hoistway vents, trolley beams, trap doors, or other provisions for removal of components of elevator equipment from elevator machine rooms.

    d. Supports for elevator machine beams, elevator deadend hitch beams, or escalator trusses.

    e. Elevations of elevator entrances at typical and non-typical floors, showing signal fixtures and control panels.

    f. Special elevator cabs and special hoistway entrances, power operated doors, space for door operation, and control system.

    g. Plan(s) and sections of escalators, loading ramps, and conveyors, showing clearances and limits.

    h. Layout of pneumatic type system showing exhausters, air lines and central runs, station locations, and details.

2. Specifications

Use modified Master or Standard Specifications if possible, supplemented by manufacturers' suggested specifications and adjusted to fit the project, as appropriate.

This summary checklist should facilitate completion of the Contract Documents Phase. On completion of this phase, written approval from the owner shall be obtained for the technical work and the AECM Contract Cost Estimate. This should be prepared prior to contract bidding, negotiation, or letting.

Although the checklists presented here and in Chapter 8 were set up for general contracting, the various phases of development can be modified to allow for fast-tracking contracting or multiple prime contracting procedures.

Chapter 10

Project Services Phase

Guidelines for project control and scheduling and procedures for the guidance of the AECM team's onsite are presented hereinafter.

But before discussion of these subjects, it should be noted that "Project Services" and "Construction Management Services" are not necessarily synonymous; construction management services must start well before the Project Services Phase; and construction management personnel's input can be most valuable during design, especially during the latter parts of the Contract Document Phase.

It should also be recognized that construction management consists only of *consultation* during the design phase, and it expands into *management, administration, and consultation* during construction.

Project services fall naturally into the latter part of the construction management services but, with the freedoms offered through the various potential forms of construction management services, definition of roles becomes most important.

It is the author's hope that conflict and misunderstanding in the future can be minimized through carefully written definitions and the use of abbreviations, such as those mentioned and described in Chapter 5.

To illustrate the foregoing points, consider five hypothetical clients—including the example C. C. University used throughout this text—and five different types of services. Study the illustration of these clients and services shown in Figure 10.1.

It is easy to see that adoption of these, or similar, identification abbreviations will facilitate description of services and will help to clear up some of the conflicts and problems that are developing between owners and design and construction professionals or organizations. These conflicts and problems result from confusion of roles and responsibilities.

The extent of services during construction must be established to fit requirements, including those of budget, timing, project, and owner. The

```
         CLIENT                         SERVICES PROVIDED

           A                            AECM

           B                            PM, PA, AECM, CGC

           C                            PM, AECM

           D                            AECMP

           E (University)               AECMM (AECM Firm)


 LEGEND:

    AECM  = Architecture, Engineering & Construction Management.
    AECMM = Architecture, Engineering, Construction Management &
            Management (University  Project).
    AECMP = Architecture, Engineering, Construction Management &
            Procurement.
    CGC   = Contractor for General Conditions.
    PA    = Project Administration.
    PM    = Project Management.
```

Figure 10.1. Examples of the types of services that might be provided to different clients by AECM teams.

extent of these services must be detailed in the Services Contract. (Refer to the sample Services Contract which is presented in Chapter 7.)

To facilitate such project services definitions in a contract, and to provide insight and general guidance, the following listings of services and sample procedures are presented. For the sake of clarity, and to again assist in establishment of guidelines for such work, the procedures given will be in a format adapted to the C. C. University Campus Project and its presumed staffing.

Services provided during construction by an AECM team would, in general, include observation, administration, and documentation; inspection and recommendation of, request for, and/or revision of plans, details, or specifications as are appropriate; and constructed quantity verification, cost estimating, Contractor's Progress Pay Estimate preparation, verification and/or approval. Other services provided by the AECM team are organization work, including jobsite meetings, scheduling—such as network layout, processing, updating and/or analysis—as well as cost control; review or analysis of subcontractor or vendor qualifications; review, evaluation, control, and approval of shop drawings, materials, and samples; labor and materials documentation; labor relations; jobsite records; progress reports and records; time and materials (force account) records; unit cost records; change orders; extra work orders; claims; and general requirement items. Other services are overall dimensions and elevations or surveys; as-built records and drawings; recommendations and communications; equal opportunity activities; insurance and bonds; disputes and arbitration; delays, suspensions of work, and termination; and time of completion. The broad scope and nature of the above listing make it obvious that efficient and expeditious services during construction will require jobsite management planning, policies, and procedures.

For the example C. C. University Campus Project, assume that services during construction are supervised by an AECM team Project Manager with a Resident Engineer Manager (Clerk of Works) and staff on jobsite. Assuming further that the AECM team Project Manager has prepared procedural guidelines for the Resident Engineer's use, these guidelines might look something like the sample (Figure 10.2) which follows.

*HYPOTHETICAL AECM CONSULTANT
DESIGN, INC.*

**GUIDELINES
FOR RESIDENT ENGINEER
DURING PROJECT SERVICES**

PROJECT

NEW CAMPUS & RELATED FACILITIES
CENTRAL CITY UNIVERSITY—OWNER
CONTRACT NUMBER 75-41

Fig. 10.2.1 Sample cover sheet for Guidelines for Resident Engineer during the Project Service Phase.

I. INTRODUCTION

The following guidelines are intended to guide the Resident Engineer and his staff in the execution of the AECM team's responsibilities for construction management, as defined in the Service Contract between the Owner, Central City University, and the hypothetical Consultant, Design, Inc.

These guidelines are not intended to supersede any codes, regulations, or contract requirements; neither are they intended to limit or abridge any of the Resident Engineer's authority or responsibility. The intent of these guidelines is to prescribe, for the purpose of general uniformity in management, the basic duties, responsibilities, policies, and procedures to be followed in performing services during construction. Numerous standard forms which have proved to be valuable have been included for convenience.

It is expected that the Resident Engineer[1] will set up his office, select his staff, and organize and manage the onsite portion of the AECM team's work.

II. FUNCTION

A. FIELD SERVICE OBJECTIVE

The *basic* objective of the AECM team's field services is to organize, coordinate, and inspect the project in order to secure completion of contracts in conformance with the plans, specifications, and approved schedules. Professionalism is required in order to insure equitable treatment of both the owner and the contractors.

B. JURISDICTION AND LIMITATIONS

Under the general direction of the Project Manager, the Resident Engineer will represent the AECM team on the project site. He will administer the contracts assigned him, organize and direct his staff assistants, and designate specific assignments for each assistant.

It is the duty of the Resident Engineer to see that the project is constructed in accordance with plans and specifications and in com-

---

[1]For simplicity and clarity, the term *Resident Engineer* is used herein to indicate the person responsible for the field administration, even though other titles, such as *Engineer-in-Charge, Project Engineer, Clerk of the Works, Field Engineer*, etc. could also be used.

The term *Contractor* as used indicates the prime contractor for each contract, including all his subcontractors and vendors.

The term *Supplier* indicates the various vendors or suppliers, including all of the subcontractors and subvendors used by the AECM team or the Owner.

pliance with all of the contract terms. The Resident Engineer has and shall exercise the authority to reject both unsatisfactory workmanship and materials. Such rejection is to be made immediately upon discovery and confirmed in writing to the contractor with references to appropriate plan or specification requirements.

Despite such rejection of materials, the Resident Engineer shall not direct the contractor to stop work, except in the case of an emergency.

Except during emergencies, the following will be acted upon only by the Project Manager, following his consideration of the Resident Engineer's recommendations:

1. Direct administration of work not assigned to the Resident Engineer, such as designing changes. Matters requiring the exchange of *formal* correspondence between the contractor, AECM team designers and consultants, and the owner will not be handled in the field unless the Project Manager directs such action.
2. Recommendation of payment to contractors or suppliers for work performed or materials furnished.
3. Authorization of performance by contractors or suppliers of extra work or changed work.
4. Authorization to omit work by contractors or suppliers.
5. Authorization of changes or substitution of materials for those specified.
6. Instructions to contractors or suppliers to start or stop work, or direction of the performance of their work.
7. Determination of policy matters or negotiation of costs. The official contact between the Owner, Central City University, and the Consultant, AECM Firm Design, Inc., is through the Project Manager. The Resident Engineer may contact the owner's personnel on routine day-to-day matters, but not on policy matters nor in regard to costs without clearing this with the Project Manager.

The Resident Engineer's staff will be tailored for the project to assist him and, in his absence or upon his specific instructions, to act for him. Personnel activities such as budgeting, hiring, firing, and changes in assignments and/or salary shall be coordinated with and approved by the Project Manager.

C. CONTACTS WITH CONTRACTORS OR SUPPLIERS

The Resident Engineer and his staff (inspectors, engineers, surveyors, etc.) shall conduct and maintain relations with the contractor(s) and/or supplier(s) in a friendly, courteous, cooperative, and businesslike fashion so that the work can be completed in the best possible manner and at a minimum cost to the contractor or supplier, in accordance with plans and specifications.

Absolute integrity is required; acceptance of gifts or favors from contractors is strictly forbidden, and excessive fraternization with key personnel of the contractor is to be avoided.

There shall be no arguments with contractors. If disagreements are unavoidable, or cannot be settled to mutual satisfaction, the matter shall be submitted immediately to the AECM team's Project Manager.

All contacts shall be administered directly with the authorized representatives of the prime contractor, *not* with his subcontractors or vendors. No direct contact will be made with manual craftsmen. In special instances, and upon prior arrangements between the Resident Engineer and the prime contractor's representative, direct contact may be made with subcontractors, but in the general case, the prime contractor will be the main contact.

D. CONTACTS WITH UTILITIES AND PUBLIC AGENCIES

The Resident Engineer will set up a procedure with the contractor(s) for contacts with utilities and public agencies. Cooperation shall be extended and a courteous reception given to representatives of utilities and public agencies. Day-by-day activities will be handled on the jobsite, but items of major importance (involving over $5,000$^{\pm}$) must be referred to or reviewed with the Project Manager.

E. CONTACTS WITH THE PUBLIC

All contacts with the public shall be handled in a courteous fashion. Specific answers to questions shall be provided or referral made to the Project Manager. Requests for public presentations and items of a similar nature should also be coordinated with the Project Manager.

The construction operations are apt to be subject to interest and criticism of the press and to large numbers of the general public. It is essential that all AECM team field employees conduct themselves in a manner that will command the respect and confidence of the public.

Complaints by the public regarding any aspect of the project's construction must receive the immediate attention of the Resident Engineer. Specific replies are to be given and, when necessary, a follow-up procedure shall be adopted to assure the adequacy of all replies.

III. CONSTRUCTION CONTRACT ADMINISTRATION

A. GENERAL

The Resident Engineer and his staff shall at all times be *thoroughly* familiar with all the provisions of the contracts which they are administering. This shall include a thorough knowledge of the plans and specifications, including all revisions, changes, and addenda.

B. PLANS AND SPECIFICATIONS

As soon as possible during design development, or at least prior to advertisement, the Resident Engineer shall be furnished an adequate supply of plans, specifications, revisions, and addenda thereto. These documents shall be studied immediately upon receipt by the Resident Engineer and his staff, and this study shall include:

1. Thorough review of plans and details and careful reading of the specifications. To the extent possible, each item in the specifications shall be compared with the requirements for the item(s) in the plans.
2. Careful checking for omissions, discrepancies, and conflicts in and between the plans and specifications.
3. Review of the designs and requirements in conjunction with the site conditions.

A compilation of any discrepancies found in the above study shall be forwarded expeditiously to the Project Manager with any pertinent comments. The objectives of this effort are to reduce the number of costly and/or time-consuming changes, extras, or clarifications, and to allow the Resident Engineer and his staff to have thorough knowledge from the outset of the plans and specifications.

C. PRECONSTRUCTION ACTION

1. Job Showings

   During contract(s) advertisement, the Resident Engineer may be called upon to assist in showing, or to show, existing site conditions. Minutes of all pertinent meetings or conversations shall be kept; such efforts will be coordinated closely with the Project Manager.

2. Receipt, Evaluation, Recommendation, and Award of Contract(s) or Subcontract(s)

   The Resident Engineer and/or members of his staff may be directed to participate in this as the Project Manager directs.

3. Early Action Letter & Preconstruction Conference

   After award of a construction contract, the Project Manager will forward to the contractor an Early Action Letter and arrange for a Preconstruction Conference. Responsible personnel of the contractor, the AECM team, and the owner, where appropriate, shall be present.

   a. At this conference, the contractor shall be oriented with respect to procedures and lines of authority for contractual, administrative, and construction matters. Procedures, methods, and forms for the following items shall be explained to the contractor:

      (1) Progress charting and scheduling (sample copies on approved format, such as bar charts or network diagrams, may be furnished).

    (2) Changes and extras.
    (3) Safety responsibilities.
    (4) Methods of submitting correspondence, shop drawings,
        and samples.
  b. This conference need not be limited to the above; other approp-
      riate matters can be introduced, such as:
    (1) Construction procedures.
    (2) Labor relations.
    (3) Special problems.
    (4) Designation of key personnel.
    (5) Equal Employment Opportunity.
    (6) Inspection procedures.
    (7) Requirements of meeting attendance.
  c. A letter of record, or minutes of the meeting, shall be prepared
      to document all items discussed at the Preconstruction Confer-
      ence, and copies furnished to all those in attendance.
4. Notice to Proceed and Commencement of Work
   A Notice to Proceed Letter will be issued to the successful contrac-
   tor(s) upon receipt of the properly executed contract documents and
   insurance and bond papers. In many cases, the date of receipt of the
   Notice to Proceed Letter is the beginning date for the construction
   calendar; or the date of receipt can fix a limiting date for commencement of the work.
       In general, the Construction Contract Documents (for Central City
   University) will require the contractor to commence work within a
   minimum number of days after acknowledgement of receipt of the
   Notice to Proceed Letter. For record purposes, the Resident Engineer will inform the Project Manager, in writing, of the date on
   which the contractor commenced work at the jobsite.

IV. CONTROL OF THE WORK

Ordinarily, the contractor will be responsible for and at liberty to select the
means or methods to be employed in performing the work. The Resident
Engineer shall make no effort to dictate a method, sequence of operations, or
direction of the work, except as required by the plans and specifications.

Only if the AECM team is specifically authorized in the plans and specifi-
cations to select or prohibit a method, and if such action is desirable, will the
Resident Engineer issue such instructions to the contractor. The Resident
Engineer will not issue written instructions without approval from the
Project Manager.

A. CONSTRUCTION SCHEDULE AND PROGRESS REPORTING

The Construction Schedule(s) shall be prepared by the contractor(s) in accordance with the contract requirements. Depending on the requirements of the contract, the format can be varied to suit project requirements.

The format of the schedule shall be reviewed by both the Resident Engineer and the Project Manager. If the schedule does not comply with the specifications, or if it is in such form that proper analysis is impossible, it should be returned to the contractor for correction and resubmission. If the schedule appears to comply with the specifications, it is to be thoroughly reviewed before it is returned to the contractor.

The Resident Engineer shall enforce the timely submission of the Construction Schedule(s) and periodical updates. These periodical updates to the Construction Schedule(s) shall show both the progress to date and the effect of any authorized changes and extra work. Time extensions shall be handled as a discretionary item by the Resident Engineer after coordinating with the Project Manager.

B. CONSTRUCTION PROGRESS CONTROL

The Resident Engineer must keep aware of *all* aspects of the progress of the project, including such items as:
1. Contractor(s)' schedule(s).
2. Status of submittals, fabrication and delivery of materials.
3. Cash flow and draw-down.
4. Progress trends of key items, such as:
    a. Excavation.
    b. Piling.
    c. Concrete work.
    d. Mechanical work.
    e. Electrical work.
    f. Finish.

If the contractor falls behind the current schedule, or if the trends in production indicate that required production rates will not be met, the Resident Engineer shall investigate and make a careful analysis of the reasons for the delay.

The Resident Engineer will keep the Project Manager informed of his findings in these matters. After such coordination, the Project Manager may himself give, or instruct the Resident Engineer to give, the contractor *written* notice of his schedule deficiency and request (*require*) the contractor to submit evidence demonstrating the manner in which the schedule will be regained.

When provided for by contract, the Project Manager may direct the contractor to increase his efforts, working hours, manpower, equipment, etc. Such action will be taken only if required, and only after a thorough review of the status of the contractor's schedule, and only after conferring with the owner, Central City University.

C. TIME EXTENSIONS

The Resident Engineer shall maintain records of work progress and delays to this progress.

All contractor-requested time extensions shall be evaluated by the Resident Engineer, and a file will be maintained in the project office.

If a contractor's request for time extension is found to be valid, he may be notified of such findings by the Project Manager. The official document granting a time extension will be a Contract Modification. If extended operating costs are to be paid the contractor, the total amount of these will be indicated in this document.

D. INSPECTIONS

It is the responsibility of the Resident Engineer to insure that the materials used and the workmanship employed are in accordance with contract documents. The Resident Engineer shall himself plan, establish staffing requirements for, and direct inspection activities.

1. Onsite Inspections

   Regular onsite inspections will be performed. Such onsite inspections will be made by personnel experienced in judging compliance with plans and specifications; these are either AECM team employees acting under the control of the Resident Engineer or employees of professional inspection subcontractors arranged for by the Resident Engineer through the Project Manager.

2. Offsite Inspections

   When contractually required, or when the Resident Engineer determines necessary, materials will be sampled, inspected, and tested at the source of fabrication prior to shipment to the project.

   The Resident Engineer will keep the Project Manager informed of all such offsite inspection requirements.

E. INSPECTION PROCEDURES, DAILY REPORTS, AND NON-COMPLIANCE NOTIFICATION

In the normal course of day-to-day work, inspectors may verbally acknowledge that the contractor's work is in apparent compliance with specifications, but the Resident Engineer will inform the contractor that such approval is not official acceptance of his work and that such official acceptance will be made at completion of the contract work.

The Resident Engineer will require each inspector to prepare daily a written report of the contractor's activities which he inspected or monitored. It will be the responsibility of each inspector to make a complete and accurate report of work and work conditions, using the "Daily Inspector's Report" form (see sample of this form illustrated in Figure 10.2.2).

In the event materials or workmanship do not comply with the specifications, it is the Resident Engineer's responsibility to notify the contractor immediately in writing of the infractions. Such notification must *not* direct the stoppage of work, but will point out that work will not be accepted if the rejected materials or workmanship are not corrected.

F. TESTING

As determined by the Resident Engineer, or as required by contract, testing will be made of materials and workmanship. The number of samples must be sufficient to permit adequate testing. The Resident Engineer will review the testing requirements of each contract and recommend to the Project Manager techniques to be employed.

Controlled testing can be performed by field personnel with field equipment. For performance testing or specification compliance tests, the subcontracting of testing will generally be necessary to obtain accurate results that will be recognized in courts of law.

Complete records will be retained of each test, and specimens will be retained if they are important to prove or disprove the passage of specification tests.

V. FISCAL ADMINISTRATION

A. GENERAL

The Resident Engineer and his staff will attempt to keep the project costs within the budget by maintaining accurate records of contractor activities and by reporting cost problems to the Project Manager in a timely manner.

B. PROGRESS PAYMENTS

Progress payments will generally be made to the contractor(s) as work progresses unless otherwise provided for in the contract. Progress pay requests will usually be prepared by the contractor(s). It will be the Resident Engineer's responsibility to review and certify the contractor(s)' progress pay requests and to recommend payment through the Project Manager to the client.

INSPECTOR'S DAILY REPORT

REPORT № _____

DATE _____

CONTRACT & № _____ SHIFT _____ WEATHER _____ TEMP _____

CONTRACTOR _____ CONTRACTOR'S REPR. _____

FEATURE(S) _____

PROGRESS OR DESCRIPTION OF WORK:

SPECIAL INSTRUCTIONS FROM SUPERVISOR:

UNUSUAL OR UNSATISFACTORY CONDITIONS:

DELAYS (length & reason):

REQUESTS FROM CONTRACTOR:

INSTRUCTION TO CONTRACTOR:

CONTRACTORS FORCES & EQUIPMENT:

REMARKS (needs, accidents, visitors, etc.):

SIGNATURE _____

Figure 10.2.2. Sample Daily Inspector's Report.

Estimates of the completed work shall be made in accordance with applicable specifications, such as those for "Measurement and Payment." It is important that all estimated percentages, estimated quantities, and final quantities be fully documented in a neat and orderly manner to substantiate the AECM team's recommendations to the client.[2]

C. COMPUTATION OF QUANTITIES AND PAYMENT PROCESSING

Generally speaking, all measurements for payment purposes will be made under the direction of the Resident Engineer rather than the contractor. It is important that all progress pay requests be estimated and final quantities be fully documented in a neat and orderly manner. Calculations are to be based upon measurements or verified observations. Remember that in some cases quantities used in progress pay estimates are subject to review by an independent auditor.

1. Computation of Quantities and Measurements
   Calculations must be prepared in a conventional manner. Often a separate page should be used for each calculation or calculated pay item. For convenience, and to standardize procedures:
   a. The specific purpose of calculations should be indicated in a title on the top of each calculation sheet.
   b. Each page of calculations is to be dated, numbered, and signed.
   c. The source of measurements is to be shown. Prior to the preparation of the first progress pay estimates, and in accordance with applicable specifications provisions, a basis for quantity measurement is to be established and progressive calculation summary sheet(s) set up. (See sample summary sheet in Figure 10.2.3.)
   d. Sketches or references are to be given to explain the calculations.
   e. Key results of calculations are to be underlined or otherwise emphasized to permit quick recognition.
   f. When items are to be paid for on a lump sum basis, the contractor(s) should be asked, at the start of construction, to break down large lump sum amounts into logical segments of appropriate amounts (approximately $1,000 or less per segment), so that progress can be judged more clearly.
   g. If items are to be paid for on the basis of drawing dimensions, the percentage completed to date will be determined by scaled or other rough measurement, or by judgment, and referenced

[2]In the situation where the AECM team makes these progress payment distributions for the client, this documentation is even more important.

PROGRESSIVE CALCULATION SUMMARY

CONTRACT & Nº   PAY REQUEST Nº

CONTRACTOR   CLOSE-OFF DATE

FEATURE OR PAY ITEM

ESTIMATED OR BID QUANTITY   LAST REPORT

                                                   THIS REPORT

                                                   TO DATE

CALCULATIONS OR SOURCES OF INFORMATION :

PREPARED BY
APPROVED

Figure 10.2.3. Sample Progressive Calculation Summary sheet.

back to drawings, indicating such things as station, section, level, or item completed.

h. For items to be paid on the basis of actual field measurement, the Resident Engineer is to establish an acceptable procedure. Often surveyed measurements are not justified until completion of each pay item. Progress quantities can be based on drawing dimensions and calculations of apparent quantities; actual though not surveyed measurements; judgment of percentage complete; or supplemental records such as truck counts, concrete mixer load slips, certified weight tickets, etc.

i. When the unit of measurement for payment is by the ton, gallon, etc., based on certified weight slips, meter readings, etc., documented records should be set up, compiled, filed, and cross referenced to support pay request quantity calculations.

2. Payment Processing Suggestions

The following suggestions for processing pay requests may be helpful:

a. Always cross check pay request quantities and monies with previous progress payments and/or bid documents.

b. Payment for changed or extra work orders should not be made until the proper documents have been fully executed.

c. Even if the contract permits, do not discontinue the withholding of retention without advising or obtaining approval from the owner.

d. If possible, determine final pay quantities for items as the items are completed, and reach agreement on these final pay quantities with the contractor(s).

e. Make sure the last progress pay request for each contractor is identified as "Final Progress Payment."

f. Withhold liquidated damages, if appropriate, from progress payment(s).

D. CHANGES TO THE WORK[3]

1. Control Requirements

The basic requirements for control of Changes to the Work are:

a. All changes to the work must be announced within the AECM team by a "Notice of Pending Change." The Project Manager will be responsible for the preparation and distribution of all Notice of Pending Change forms. The Resident Engineer will coordinate with the Project Manager in this work. A description

[3]"Changes," in this context, include changed, new, and/or extra work.

of the procedures to be followed can be found in Section F, "Notice of Pending Changes," of these instructions.

b. When changes involve adjustments to contract prices or contract time, a "Contract Modification" (Change Order) must be prepared. See Section G, "Directives, Proposals, Change Orders, and Contract Modifications," of these instructions for a description of procedures pertaining to contract modifications.

c. All changes must be made or authorized in writing, and the effect upon contract cost and time stated, except in cases of emergency changes, as described in Item 2, below, or as determined appropriate and desirable by the Resident Engineer or Project Manager.

2. Emergency Changes

Emergency changes are those for which the time of processing would be too long and would result in excessive costs to the owner. In such cases, the Resident Engineer is to take the following steps:

a. Review the change with the contractor to determine the cost/time adjustment. If possible, obtain approximate cost/time or guaranteed maximum cost/time figures.

b. Review the change and circumstances with the Project Manager, who will proceed with obtaining approval, as appropriate, from the owner.[4]

c. Authorizing Emergency Changes

(1) If the estimated (or, preferably, the maximum) cost is less than $3,000, the Resident Engineer may authorize the work.

(2) If the estimated (or, preferably, the guaranteed maximum cost) is between $3,000 and $10,000, the Resident Engineer will obtain telephone approval from the AECM team Project Manager (or from the Controller of C. C. University, Owner).

(3) In the event the estimated or guaranteed cost exceeds $10,000, approval will be obtained by the AECM team Project Manager from the President of Central City University, Owner.

E. REQUESTS FOR CHANGES TO THE WORK

Requests for changes can be made by the Owner (C. C. University's

---

[4]It is highly desirable for the AECM team's Project Manager to establish guidelines for emergency changes authorization with the owner. In this case, these might appear as in Item c, "Authorizing Emergency Changes," which follows. Provisions such as these could also be included in the AECM team's contract with the owner.

designees), AECM team members, the contractor(s), outside consultants, or by regulating agencies.

Requests for changes should be prepared in writing and be as complete as possible. If initiated by members of the AECM team, a Notice of Pending Change form may be employed.

In the case of emergency changes, only a directive to the contractor(s) need be prepared.

1. Requests by Client (Owner)

    Desires of the Client (C. C. University and its designees) shall be reviewed by the Project Manager and Resident Engineer. These desires may be discussed and reviewed within the AECM team. Results of this review will be transmitted to the client with the AECM team's recommendations. If the client elects to proceed with the proposed change, the AECM team will prepare appropriate drawings and/or specification changes, cost estimates, and proposal requests. The Project Manager and Resident Engineer will coordinate in making the request for the proposal of the contractor(s).

2. Requests by Contractor(s)

    The Resident Engineer is to establish a procedure with the contractor(s) for the presentation of these requests. In general, requests by the General or Prime Contractor(s), including those originating with subcontractors, will be submitted by the *contractor* in writing to the Resident Engineer and/or Project Manager.

    Except in an emergency, requests should be in letter form and shall include a firm proposal covering costs and time. Such proposals may indicate the latest date for acceptance.

    The Resident Engineer shall transfer all Contractor Change Requests to a Notice of Pending Change Form and forward them to the Project Manager with his recommendations and comments. (See Section F, "Notice of Pending Change," of these instructions.)

    Emergency requests by the contractor should, if possible, be prepared similarly and should contain as much information as is available, including an estimate of extra costs and/or delay which may be incurred if a decision is delayed beyond a stipulated date. The Resident Engineer will review the requests and forward or deliver them to the Project Manager with comments and recommendations. If the Resident Engineer determines it appropriate in an emergency, he may elect not to put the contractor's change request on a Notice of Pending Change Form.

    The Project Manager, or, in his absence, the Resident Engineer, will present the emergency request to the owner. The contractor should be advised on the acceptability of his requests to the owner through the Project Manager or Resident Engineer. (See Section D.2.c, "Authorizing Emergency Changes," which offers an obviously desirable solution.)

3. Requests by AECM Team or Outside Consultant

Change requests originating from AECM team members (design and other) or from outside consultants are to be prepared on Notice of Pending Change forms and transmitted to the Project Manager, who will insure that the Resident Engineer receives them for review.

After this review, the AECM team's comments and recommendations will be presented by the Project Manager to the client with a request for approval. If approved, the Project Manager will have a Request for Proposal prepared. The Resident Engineer will participate in this; no Request for Proposal will be sent to the contractor without the Resident's knowledge or authorization.

4. Requests by Regulating Agencies

Federal, state, and local agencies may request changes to the drawings and specifications because of building codes, local laws and regulations, or agencies' desires.

Generally, all designs and documents will have been reviewed and approved by such agencies in a previous phase of the AECM team's work. Nevertheless, problems can develop, and it is the Resident Engineer's responsibility to establish onsite procedures to control such situations. This type of change request will be processed in a similar manner to those mentioned in Item 3, above, the exception being that the Resident Engineer is to translate the request to a Pending Change Notice form.

F.   NOTICE OF PENDING CHANGE

For uniformity, a "Notice of Pending Change" (NPC) form, which is standard and preprinted, will be used on this project. (See sample NPC form in Figure 10.2.4.)

NPC's are letter-size forms containing sketches with notes or descriptions; these are to be used by authorized personnel to identify proposed modifications to existing contract documents which could result in new or changed contract work. NPC's can serve as a transmittal for design changes to contract drawings; in such cases, changed drawings will be marked "preliminary" and referenced to the attached NPC.

NPC's which would make changes to the basic design will only be made by or with the approval of the responsible AECM team designer.

Field originated changes will be made by or under the direction of the Resident Engineer. He will insure that the NPC is signed, dated, and forwarded to the Project Manager. Further, if the change is due to

Page _____ of _____ .

NOTICE OF PENDING CHANGE

NPC Nº _____   CONTRACT & Nº _____
PREPARED BY_____   DRAWING(S) Nº _____
DATE _____   SPEC. REF _____
                                _____
                                _____

NECESSITY FOR CHANGE_____

_____

_____

DESCRIPTION OF CHANGE:

1. Prepared on pre-printed form by originator.

2. Submit with preliminary cost estimate.

3. Submit form with estimate and one copy of each.

RECOMMENDED BY   DATE   CONTRACTOR

APPROVED BY   RECEIVED BY   DATE

Figure 10.2.2.4. Sample Notice of Pending Change.

conflicting field conditions or interferences, he will insure that an adequate description is provided.

Upon receipt of a NPC form from whatever source, the Project Manager will insure that proper distribution is made within the AECM team. Two copies of all NPC's, prepared by an AECM team member (designer or other) or by outside consultants, will be furnished to the Resident Engineer for review. If the change cannot or should not be made, he will notify the Project Manager immediately. The Project Manager will insure that all NPC's are properly processed and that approvals, if appropriate, are obtained.

An alpha-numerical control system of identification for NPC's will be set up by the Project Manager. In general, this consists of a contract identification number, an alphabetical identification of the source of the NPC, and a sequential identification number or numbers.

NPC's are designed to aid communication within the AECM team and may serve as a basis for documents which change or request a proposal for the changed work described next. The contractor(s) need never see a Notice of Pending Change.

G.   DIRECTIVES, PROPOSALS, CHANGE ORDERS, AND CONTRACT MODIFICATIONS

Because of the numerous financial and other problems that can become involved in making changes to the work or the contract, the Project Manager and Resident Engineer will coordinate their efforts in this regard, and the client will be kept fully informed by the Project Manager.[5]

1. Change-Related Documents

Often, the Contractor-Owner contract will prescribe change procedures and documents. Change-related documents in AECM work can include the following:

a.   Letters of Instruction or Directives (Written Orders)

Written Orders will be handled as indicated by Section VI.B, "Written Communications," presented later in this chapter. It is the responsibility of the Resident Engineer to establish field procedures to control written instructions by his staff to the contractor(s); however, excessive written correspondence between even the most senior of his staff and the contractor(s)

---

[5]In general, it is contemplated that the specifications for the work will establish limits on the amount(s) of money the Resident Engineer and/or Project Manager can authorize or commit for the client to the contractor. This authority may also be spelled out in the AECM Team-Client contract. If not, it is suggested that the Project Manager work out a mutual agreement with the client and obtain a written authority to at least make a commitment in cases of emergency.

should be discouraged. Further, it is noted that *no* direct written correspondence between the field staff and subcontractor(s) is to be allowed.

  b. Field Directives and/or Field Orders
      Field Directives and Field Order procedures will be worked out between the AECM team (Resident Engineer and Project Manager) and the owner prior to the preconstruction conference. The procedures adopted will be adhered to. Generally, and if possible, a standard form requiring three authorizing signatures will be employed. (See the sample Field Directive in Figure 10.2.5.)
      It will be the responsibility of the Resident Engineer to obtain statements of authority from the contractor related to an ability to commit the contractor (especially in the cases of corporations or joint ventures) in matters of changes and costs.

  c. Contract Change Proposal Requests
      Contract Change Proposal Requests (see Figure 10.2.6) will be sent by transmittal to the contractor(s) under the signature of the Project Manager. Nevertheless, the Resident Engineer is to become thoroughly familiar with these on or before issuance and be ready to assist the Project Manager or the contractor(s) with difficulties related to these requests for proposal.

  d. Contract Change Orders
      Contract Change Orders will be issued by the Project Manager, but the Resident Engineer will provide any requested assistance with these that he can. (See Figure 10.2.7.)

  e. Contract Modifications
      Contract Modifications will also be issued by the Project Manager, with assistance requested from the Resident Engineer as needed. (See Figures 10.2.8.A and B.)
      In general, all forms of changes or extras are to be ultimately included in a Contract Modification. Prior to such a modification, however, it is the joint responsibility of the Resident Engineer and the Project Manager to assure that a fair and reasonable or equitable price adjustment is reached with the contractor.

2. Estimating for Contract Change Negotiations
   In most cases, an estimate prepared by the AECM team's estimator will be used by the Resident Engineer as the basis for negotiations with the contractor(s). Under certain circumstances, such as having first-hand knowledge, retrospective pricing, etc., the Resident Engineer may prepare this estimate; however, it must be checked and

Engineers

400 Building 400 108 th Avenue N E - Bellevue, Washington 98004
Phone   (206)   455·9230   or   232·1190

FIELD DIRECTIVE NO. ____

To: _____   Date: _____
    _____   Contract No. _____ _ __
Attn: _____   Description: _____

Work to be Performed: _____

_____

_____

_____

Reason for Modification:_____

_____

Work under this Field Directive Modification will be accomplished:

☐ With no change to the contract price.

☐ For a lump sum increase/decrease of _____
   to the contract price.

☐ On a unit price basis with a maximum allowable increase/
   decrease of _____

☐ With work to start immediately and determination of any
   change to the contract price and/or completion date to
   be negotiated. Negotiations to be complete by _____
   (within 7 working days).

☐ With no change in contract completion date.

☐ With contract completion date extended/decreased _____
   calendar days.

Performance of this Field Directive is authorized by signature of
the Owner's Representative, below.

Accepted by:_____
            (Contractor's Rep.) (Title)   (Date)

Approved by:_____
            (AECM Team's Rep.)   (Title)   (Date)

Authorized by:_____
              (Owner's Rep.)   (Title)   (Date)

Figure 10.2.5. Sample Field Directive. (form courtesy Constructioneering Northwest Inc.)

400   Building, 400 108th Avenue N.E. - Bellevue, Washington   98004
Phone (206) 455-9230 or 232-1190

Engineers   CONTRACT CHANGE PROPOSAL NO. _____

Contract No._____   Description:_____

REQUEST FOR PROPOSAL

To Contractor: _____(Contractor's Name & Address)_____

In connection with your contract with _____(Owner)_____
dated _____, please furnish your proposal for performing the
changes outlined below and/or detailed on the attachments referred to:

AECM Firm Y _____ By_____ Date_____

PROPOSAL

To

We propose to perform all changes described in the above request for a
total change to the contract sum (not including State Sales Tax) of:

ADDITION/DEDUCTION of _____$_____

We have attached hereto Cost Estimate Detail Sheets Nos. _____
The foregoing amount covers all direct and indirect costs related to this
change and to the effect of the change on the remainder of the project.
All other provisions of the contract remain in full force and effect.
We request the completion date be extended _____ calendar days because of
this work.

_____ By_____ Date_____
       (Contractor)

RECOMMENDATION

To: _____

We have carefully examined the foregoing proposal and find it to be in
order and the cost reasonable. We, therefore, recommend its acceptance.

_____ By_____ Date_____

AUTHORIZATION

The foregoing proposal is accepted and authorizes the performance of the
changes specified. This instrument, when signed below, constitutes
authority to proceed with the above work. A formal Change Order in this
amount will follow. Billings cannot be honored for this change until
issuance of the formal Change Order.
AUTHORIZED BY:

_____ By_____ Date_____
       (Owner)

Figure 10.2.6. Sample Contract Change Proposal Request. (form courtesy
Constructioneering Northwest Inc.)

400 Building, 400 108th Avenue N.E. - Bellevue, Washington 98004
Phone (206) 455-9230 or 232-1190

Engineers

CONTRACT CHANGE ORDER NO. _____

Contract No. _____ Description: _____

REQUEST FOR PROPOSAL

To Contractor: _____ *(Contractor's Name & Address)* _____

In connection with your contract with _____ *(Owner)* _____
dated _____, please furnish your proposal for performing the
changes outlined below and/or detailed on the attachments referred to:

_____ By_____ Date_____

PROPOSAL

To _____

We propose to perform all changes described in the above request for a
total change to the contract sum (not including State Sales Tax) of:

ADDITION/DEDUCTION of _____ $_____

We have attached hereto Cost Estimate Detail Sheets Nos._____
The foregoing amount covers everything required in connection with this
change. All other provisions of the contract remain in full force and
effect. We understand that no invoices incorporating the amount of this
change will be honored prior to authorization or prior to the performance of the work specified.

_____ By_____ Date_____
(Contractor)

RECOMMENDATION

To:_____

We have carefully examined the foregoing proposal and find it to be in
order and the cost reasonable. We, therefore, recommend its acceptance.
Following is a summary of the contract amount:

Original Contract Sum: $_____   Previous Total: $_____

Previous Additions: $_____   This Change (ADD): $_____
                                                  (DED): ($_____)

Previous Deductions: $_____   New Total: $_____

_____ By_____ Date_____

AUTHORIZATION

_____ hereby accepts the foregoing proposal and
authorizes the performance of the changes specified. This instrument
constitutes a change order to the contract only when authorizing signature is affixed below.

AUTHORIZED BY:

_____ By_____ Date_____
(Owner)

Figure 10.2.7. Sample Contract Change Order. (form courtesy
Constructioneering Northwest Inc.)

140

DRAFT CONTRACT MODIFICATION

Contractor: _____   Contract No. _____
            _____   Description: _____
            _____   _____

                                       Modification No. _____
                                       Title:_____
Contract Modification: _____

  ☐ Change within Scope of Work.

  ☐ Issued because of Supplementary Agreement.

  ☐ Other (Explain)_____
     _____

Notice of Pending Change(s) for this Modification:
_____
                  DESCRIPTION OF CHANGE(S)

Contract Time:	Amt. this Modification: $_____
The completion date, as well as all other terms, covenants and conditions of the Contract, except as modified by this and previous modifications, if any, remains in full force and effect.	From previous Modifications____ Total Modifications to date____ Original Contract Amount_____ Revised total of con- tract, including this Modification_____

(over)

Figure 10.2.8.A. Sample draft of a Contract Modification (front side of form).

DRAFT CONTRACT MODIFICATION - PAGE TWO

ACCEPTANCE BY CONTRACTOR:

Accepted: (Name of Contractor)

By: _____ (Typed Name) _____

Title: _____ (Typed Title) _____

Signature: _____

Date: _____

* APPROVAL/RECOMMENDATION BY AECM FIRM

Approved: _____

By: _____ (Typed Name) _____

Title: _____ (Typed Title) _____

Signature: _____

Date: _____

AUTHORIZATION/ACCEPTANCE BY OWNER:

Authorized: _____

By: _____ (Typed Name) _____

Title: _____ (Typed Title) _____

Signature: _____

Date: _____

*NOTE: The AECM team's authority to approve or accept should be
established in the Owner-AECM Team and Owner-Contractor contracts.
This signature may be eliminated if not required by contracts,
codes or statutes.

Figure 10.2.8.B. Sample draft of reverse side of a Contract Modification form.

approved by the Project Manager before use in negotiations with the contractor(s). The estimator should strive to obtain lump sum agreements, although other forms of adjustment of price for changes and extras may be used.

    a. Work Performed for a Lump Sum

        This is the preferred form of agreement and payment for extras and changes, especially because it can be made inclusive of many work items and because this type of payment agreement tends to simplify field accounting and reporting.

    b. Work Performed at Unit Prices

        If the contract contains unit prices and extra or changed work can be performed for this original bid unit price, this form of price adjustment is satisfactory.

        If new unit prices are established, this form of adjustment is also acceptable. The disadvantages of this type of cost agreement are the requirements for measurement, uncertainty of final price to the client, and the additional field accounting and reporting required. Nevertheless, the Resident Engineer may find this type of agreement of value.

    c. Work Performed on Force Account or Cost Plus Basis

        Although this form of compensation may be thought to destroy the contractor's performance incentive, this is not necessarily so. If extra or change work is, or can be done on this type of basis, the Resident Engineer will attempt to insure equity of such an arrangement and will insist on adequate performance by the contractor's personnel.

        Work commencement, progress, and completion will be documented adequately and noted on Daily Progress Reports and in diaries. The Resident Engineer will appoint a specific individual to report on materials used and time for labor and equipment. The Resident Engineer will insure that the contractor's force account or cost plus billings agree with the AECM team's records, including inspectors' reports. All inspectors' and other reports will be kept on file in the Resident Engineer's office.

H. PROCESSING OF CONTRACT MODIFICATIONS AND FINDING OF FACTS

In most cases, the Resident Engineer, after reaching an equitable tentative agreement with the contractor, is to prepare a "Finding of Facts" report and recommendation for the Project Manager to process. Ideally, the recommendation will include a draft of the Contract Modification.

(See the sample forms entitled "Finding of Facts for Pending Contract Modification" shown in Figures 10.2.9.A and B.)

The Finding of Facts is to contain sufficient information relative to the changed or extra work to justify the amount of compensation agreed upon. If the change is design oriented, or if the Resident Engineer determines it appropriate, he may seek help in the preparation of the Finding of Facts from members of the AECM team through the Project Manager. Information such as the following is to be included:

1. Origin of the change.
2. Necessity for the change.
3. Description of the change.
4. References to the changed work to drawings and/or specifications.
5. Items involved in the change, together with itemized negotiated amounts, if possible, and an estimate or statement of the total amount of monies involved in the change.
6. A statement or estimate of the change in or effect on contract time.
7. The Resident Engineer's complete comments regarding the change, including a description of the circumstances resulting in the tentative negotiated amount and his recommendations or suggestions.
8. Attachments:
   a. Copies of all correspondence from the contractor that is pertinent to the change, including *all* proposal letters.
   b. Copies of the AECM team's estimate of costs.
   c. Drawings or sketches related to the change but *not* to be part of the change or Contract Modification.
   d. Complete records of negotiations and results.
   e. Other appropriate documents necessary to present a complete Finding of Facts and recommendation.

In the event an equitable adjustment or satisfactory price cannot be agreed upon between the Resident Engineer and the contractor, other avenues of price negotiation will be pursued by the Project Manager. If possible, cost plus and force account agreements will be entered into, rather than authorization of work without a total price agreement.[6]

---

[6]This procedure may seem to place tremendous authority and confidence in the AECM team and its Resident Engineer relative to extras and changes, at least to some owners; and alternative procedures may and should be developed for some projects. Nevertheless, this guideline, which is only suggestive, should help in organizing thought and establishing procedures to fit the circumstances of the owner and the project. This is a big plus for the AECM team approach.

If desired, instead of employing the type of forms previously shown and discussed in these guidelines, other forms may be designed and used. The Resident Engineer and Project Manager will determine, based on the contract and project requirements, which forms shall be used.

Engineers

400 Building, 400 108th Avenue N.E. – Bellevue, Washington 98004
Phone (206) 455-9230 or 232-1190

FINDING OF FACTS FOR PENDING CONTRACT MODIFICATION

Contract No. _____

Description: _____

Notice of Pending Change No. _____

Priority: (Explain) _____

_____

_____

Necessity for Modification: (Explain) _____

_____

_____

Description of Work: _____

_____

_____

_____

Drawings: _____   Spec. Reference: (or Revisions)

_____   _____

_____   _____

Estimated Cost: _____

Recommended Basis of Payment: _____

Estimated Change Contract Time: _____

Resident Engineer's Recommendation: _____

_____

_____

_____
(Signature)
Resident Engineer

Figure 10.2.9.A. Sample Finding of Facts for Pending Contract Modification form (front page).

Constructioneering
Northwest
Inc.

FINDING OF FACTS - PAGE TWO

Resident Engineer's Comments: _____

_____

_____

_____

_____

                              _____   _____
                              (Signature)   (Date)

Project Manager's Analysis & Recommendation: _____

_____

_____

_____

_____

                              _____   _____
                              (Signature)   (Date)

Attachments: (Check appropriate squares.)

☐ Contractor's Proposal (if obtained).

☐ AECM Team's Estimate Recap.

☐ Drawings.

☐ Sketches.

☐ Notes.

☐ Record of Discussions or Negotiations with Contractor.

☐ Other Documents (List).

    _____

    _____

Figure 10.2.9.B. Sample Finding of Facts for Pending Contract Modification form (reverse side).

VI. GENERAL FIELD OFFICE ADMINISTRATION

The Resident Engineer is responsible for overall field office administration, and he shall manage and expeditiously handle this effort with and through his staff by delegating authority to and indicating the responsibilities of his staff.

A. UNWRITTEN COMMUNICATIONS

All communications which are or could be important to the administration of the contract must be substantiated by permanent records, such as correspondence, minutes of meetings, and written notes of verbal conversations. Summaries of all important unwritten communications should be filed or confirmed. Notes should be taken at conferences and during telephone calls and discussions; they should indicate the date, location, parties involved, and important topics discussed.

B. WRITTEN COMMUNICATIONS

As field representative for the AECM team, the Resident Engineer will have the basic responsibility for field communications with the contractor, subject to the limitations imposed by the contract and those imposed by and worked out or coordinated with the Project Manager.

C. OUTGOING CORRESPONDENCE

Generally, all field correspondence to the contractor, as well as to other parties, should be handled by the Resident Engineer, except for memos and transmittals of a routine nature. (See the sample outgoing letter in Figure 10.2.10.)

Outgoing written correspondence should be similar to the sample AECM letter, and shall be governed by or in accord with the following general rules:
1. The letter must be addressed to the proper party (generally by firm name, and to the attention of a responsible or *senior* member of the firm), with the designated number of copies.
2. Distribution procedures shall be set up by the Resident Engineer or Project Manager, and file copies shall be made in accordance with a color code similar to the one suggested in Item 4, below, of these rules.
3. The distribution of carbon copies shall generally be indicated on the original and all carbon copies. Use the letters "cc:" followed by the parties' names at the bottom of the letter after the signature to indicate which parties received a copy of the letter. If it is not desirable to show, on the original copy of the letter, the name of any party to whom a carbon copy is being sent, then a "blind carbon copy" may be sent. In this case, the letters "bcc:" followed by the

DESIGN INC.

10 MAIN STREET

CENTRAL CITY, CALIFORNIA 99999

Letter Serial No. 1

(Date)

University
10 Campus Road
Central City, California 99999

Attention: Mr. A. B. Cole, President

SUBJECT: UNIVERSITY - CAMPUS PROJECT
            SAMPLE OUTGOING LETTER
            FILE NO. _____

Gentlemen:

To the extent practicable, the subject and file number should be designated for all correspondence by the originator, unless they are obvious
or can be conveniently left to a responsible secretary. The project
name, in this case, "University Campus Project," should be shown on
the first subject line of <u>any</u> correspondence pertaining to the project.

A serial number system shall be established by the Resident Engineer,
and separate serial number systems shall be maintained for each contract
or subcontract.

The salutation, complimentary close, signature block, initials of the
originator and typist, and distribution, with or without enclosures,
shall be arranged similar to those shown hereon. All correspondence is
to be centered and well balanced. The typing shall be single-spaced,
and paragraphs are to be blocked. The margins shall depend on the length
of the letter, but preferably shall not be less than 1¼ inches on the
left side and one inch on the right side.

Carbon copies shall appear as indicated. The names of persons receiving
blind carbon copies shall appear only on their copies and on the appropriate colored tissue copies.

Very truly yours,

J_____ G_____
Project Manager

JG:mlh
Enclosures (3)
cc: P.D.Q. (w/encls.)
     N.O.P. (w/o encls.)

Figure 10.2.10. Sample AECM Team Outgoing Letter.

party's name would be shown on all AECM file copies except that destined for the Reading File.

4. Tissue file copies of correspondence shall be prepared in accordance with the following color code:

    Pink: Copies for the master files.

    Yellow: Copies for the chronological and reading file (routing copy).

    Blue: Copies for the originator's file.

Generally, two parallel files will be maintained, one onsite and the other in the Project Manager's offices. Specifics of this are to be worked out by the Resident Engineer and Project Manager.

5. The subject line shall include an adequate subject description and may, at the Resident Engineer's discretion, include contract identification and/or file number.

6. Each letter should be confined, so far as practical, to one subject.

7. If the authorized signator (Project Manager, Resident Engineer, or other) is temporarily absent, a person should be designated who will sign his own name and insert the work "for" in front of the signator's name.

8. When replying to a letter with more than one subject, a separate letter normally should be prepared for each subject.

9. When replying point by point to an incoming letter, answer *all* points raised, preferably in the order in which they appear. In cases where incomplete answers are made, indicate why complete answers cannot be provided.

A sample outgoing memorandum is provided in Figure 10.2.11. The same general rules, above, apply to the use of the memorandum form of correspondence as they do to the more formal letter, except that the memorandum is used for the more routine matters.

Also, a sample copy of the Transmittal form is provided in Figure 10.2.12.

D. INCOMING CORRESPONDENCE

The following general principles and rules will apply:

1. To the extent practical, originators of incoming correspondence are to be requested to supply the field office with an original and one copy of all written correspondence. The original will be used for routing, preparation of answers, and permanent filing, while the copy will be filed chronologically.

2. The Resident Engineer's secretary, or another person designated by him, will open all field office mail except that marked "personal" or "confidential," date stamp all incoming correspondence, assign file number, log in the correspondence, if appropriate, route stamp the original, and route it to and/or through the Resident Engineer.

DESIGN INC.

10 MAIN STREET

CENTRAL CITY, CALIFORNIA 99999

TO: A. B. Cole - University   MEMO SERIAL NO. 2

FROM: J _____ G _____ - AECM Firm   DATE: July 11, 1975

SUBJECT: UNIVERSITY - CAMPUS PROJECT
           CONTRACT NO. 1234
           SMART CONSTRUCTION, INC. (ROADS)
           ADDITIONAL TIME REQUEST
           FILE NO. 611.3

On Thursday, July 10, 1975, B. E. Smart, Superintendent for Smart Construction Company, requested a time extension of eight working days
because of a delay in delivery of owner-furnished materials and the
changes in Drawing No. S-41/Rev. 1, dated July 1, 1975, which required
a modification to the concrete form.

The owner-furnished materials which were delayed in this case are imbed
plates. Although delivery of these materials was scheduled for July 1
in the specifications, actual delivery was not made until July 5, 1975.

On July 1, 1975, Drawing No. S-41/Rev. 1 was received, affecting dimensions on concrete wall forms and Column A-2, and changing location of
imbeds. Since form fabrication was already in process at the time this
revised drawing was received, incorporation of these modifications did
delay the work until July 10, 1975.

I recommend that this additional time request be accepted since, in my
estimation, the delays in delivery and fabrication changes which occurred
concurrently did delay this critical work approximately eight working
days.

                              _____(signature)_____
                              J _____ G _____
                              Project Manager

JG:mlh

cc: S.C.I.
     P.D.Q.
     N.O.P.

Figure 10.2.11. Sample AECM Team Outgoing Memorandum.

DESIGN INC.

TRANSMITTAL NO. _____

TO: _____   DATE: _____
    _____   FILE NO.:_____
    _____   SUBJECT:_____

ATTENTION:_____   _____

Gentlemen:

We are sending you ☐ Herewith ☐ Under separate cover, via
_____ the following items:

☐ Reproducibles ☐ Prints ☐ Samples ☐ Letter ☐ Memo
☐ _____

No.	Qty.	Dated	Description

These are transmitted for the purposes checked below:

☐ For approval.   ☐ Approved as submitted.
☐ For your use.   ☐ Approved as noted.
☐ As requested.   ☐ Returned for corrections.
☐ For review & comment.   ☐ Other_____
☐ Disapproved/Resubmit.   _____

REMARKS:_____
_____
_____

COPIES TO:_____

          _____   Very truly yours,

                                      _____
                                          (Signature)
                                      (Typed Name and Title) AECM Firm

Figure 10.2.12. Sample Transmittal Form.

3. If both an original and copy are received, the copy shall be annotated with the proper file number, logged in if appropriate, and filed as soon as possible in the chronological file. In the event no copy was received, a duplicate is to be made and filed in the chronological file.

4. The Resident Engineer and/or the addressee will designate any necessary action or information routing.

5. Each person designated on the routing stamp or routing slip will promptly process the letter and deliver it to the next person on the routing list. To speed up this process, automatic stamps imprinted with the name of the individual and a changeable date may be used advantageously by those persons processing these materials in any quantity.

6. Prior to initialing or stamping, all incoming correspondence shall be read carefully.

7. Correspondence requiring replies or action shall be handled promptly. To avoid problems, mistakes, and delays in answering correspondence (since additional time may be required for study, investigation, etc.), an effective suspense system shall be established.

8. If there has to be a delay in providing an answer to a letter, an acknowledgement of the letter should be issued to the sender explaining the cause of the delay and indicating when he may expect a reply.

9. Complete files of all incoming correspondence shall be kept at the field office.

E. CORRESPONDENCE DISTRIBUTION AND FILING SYSTEMS

Procedures for distribution of correspondence will be worked out between the Resident Engineer and Project Manager. This distribution can be indicated in tabular or chart form similar to Figure 10.2.13, which follows. This is representative of the probable system.

CORRESPONDENCE	FIELD OFFICE	PROJECT MANAGER'S OFFICE	OUTSIDE CONSULTANTS	CONTRACTOR CLIENT
To Contractor	C	C		O,C
From Contractor	O,C	C		
To Client	C	C	CR	O,C
From Client	CR	O,C		C
To Outside Consultant	C	C	O,C	
From Outside Consultant	CR	O,C	C	

O = Original;   C = Copy;   CR = Copy Requested

Figure 10.2.13. Correspondence distribution chart.

Modifications and enlargements of this should be worked out. Direct correspondence between client and contractor is to be discouraged, restricted, and/or prohibited, as applicable.

Either of the filing systems or modifications thereof that follow may be selected by the Resident Engineer (with approval of the Project Manager). System A is most adaptable to building construction where System B is more complicated. Nevertheless, System B is effective for multi-contract and multi-project filing.

1. Filing System A (Building Construction)

All correspondence, contractual documents, records, drawings, reports, notices, and other documents and data accumulated in the field office shall be filed as follows:

FILING SYSTEM A—FILE INDEX

File Number		Description
A		Correspondence
1A		Prime or General Contractor(s)
	1Aa	Shoring and Underpinning
	1Ab	Excavation
	1Ac	Foundations
	1Ad	Vertical Transportation
	1Ae	Electrical
	1Af	Mechanical (as required)
2A		Design A & E (as required)
	2Aa	Architectural
	2Ab	Structural
	2Ac	Soils Investigation
	2Ad	Vertical Transportation
	2Ae	Electrical
	2Af	Mechanical
	2Ag	Surveying
	2Ah	Adjacent Property
	2Ai	Testing (as required)
3A		Client (as required)
4A		Regulating Agencies
5A		Conference Notes
	5Aa	Prime or General Contractor(s)
	5Ab	AECM Team
	5Ac	Client (as required)
6A		General & Miscellaneous Correspondence

7A			Subcontractors
B			Contracts
1B			Client and Contractor(s)
2B			Client and AECM Team
3B			Client and Consultants
4B			AECM Team and Outside Consultants
	4Ba		Soils
	4Bb		Surveying
	4Bc		Planning (as required)
C			Change and Extra Work Orders
1C			Design Changes
2C			Design Extras
3C			Construction Changes
	3Ca		Value Engineering
	3Cb		Change Conditions
	3Cc		Emergency Changes
	3Cd		Field Orders
4C			Construction Extras
5C			Potential Claims
6C			Claims
7C			Delays and Time Extensions
D			Budget and Schedules
1D			Budget
2D			Schedules
	2Da		Networks
	2Db		Printouts (as required)
E			Estimates and Bid Analysis
F			Progress Payments
1F			Contractor(s) (as required)
2F			Invoices to Client and Client Payments
	2Fa		Testing
	2Fb		Reproductions
	2Fc		Equipment (as required)
G			Reports
1G			Weekly Progress Reports
2G			Monthly Progress Reports
3G			Manhour Reports
4G			Cost Reports
5G			Labor Relations
6G			Material & Status Reports (as required)
7G			Progress Photos

H			Inspection and Control Tests
	1H		Daily Inspection Reports
	2H		Contractor(s) Daily Reports
	3H		Soils
	4H		Foundations, Piles, etc.
	5H		Reinforcing
	6H		Concrete
	7H		Structural Steel
	8H		Window Wall
	9H		Electrical
	10H		Mechanical
		10Ha	Plumbing
		10Hb	HVACol
		10Hc	Sprinklers
	11H		Finishes (as required)
I			Safety
J			Visitor Register
K			Shop Drawings
	1K		Logs (as required)
L			Tenant Improvements and Interiors (as required)
M			Miscellaneous Correspondence and Mail

File System A can and should be expanded as required to fit the requirements of the project.

2. File System B (Multi-Contract/Multi-Project)

The general concepts of this file system are:

a. Subjects of a construction nature are to be filed within a 98-number file span assigned to that specific contract. For example:

(First Contract)
1-600 to 1-698

If a second contract exists, it would be preceded by the number 2 and construction filing would be by subjects numbered 600 through 698.

b. Subjects of a construction nature, but which cannot be identified with a specific contract, could be filed with File Number 699 or subdivisions thereof.

c. Chronological files or logs will be maintained for use as "Reading Files" and as a locater for numbered subject file correspondence.

      d. On receipt of correspondence from an outside source, the sub-
         ject file number will be noted on both the original and the
         chronological file copy.

This system is intended to facilitate maintenance of parallel records between the Project Manager's and the Resident Engineer's offices. Nevertheless, since the Project Manager may be using other number groups, such as:

100-198   Planning
200-298   Design

there is, or can be a problem in maintenance of the parallelism of these files
unless these two sets of files are cross checked periodically. The field office
files will only use numbers 600-699 (or as assigned).

FILING SYSTEM B—FILE INDEX

Chronological Files
        Chrono—Outgoing to Contractor(s)
        Chrono—Outgoing to Project Manager
        Chrono—Outgoing to Public Agencies
        Chrono—Outgoing to Utilities
        Chrono—Outgoing to All Others
        Chrono—Incoming from Contractor(s)
        Chrono—Incoming from Project Manager
        Chrono—Incoming from Public Agencies
        Chrono—Incoming from Utilities
        Chrono—Incoming from All Others

Number	Description
600	CONTRACT MATTERS
600.1	Preconstruction Activities
600.2	Preconstruction Surveys
600.3	Preconstruction Photos (as required)
601	Conformed Copy of Contract Documents (as appropriate)
601.1	Contractor-Subcontractor Listings
601.2	Contractor-Subcontractor Management Information
601.3	Contractor-Subcontractor Notices
602	Change and Extra Work Orders
602.1	Changes
602.2	Extras
602.3	Field Directives (as required)
603	Progress Invoices & Progress Pay Estimates
603.1	Progress Payment Back-up Calculations

603.2	Progress Photos
603.3	Forecasts (as required)
604	Meetings and Phone Calls
605	Contractor-Subcontractor Personnel
606	Submittals
606.1	Submittal Procedures, Requests, Schedules, and Special Problems
606.2	Site Work (Expand as required; use numerical code from CSI or Specification sections as appropriate. This note is typical for all subsequent 606.X numbers.)
606.3	Concrete
606.4	Masonry
606.5	Metals
606.6	Carpentry
606.7	Moisture Control
606.8	Doors, Windows, and Glass
606.9	Finishes
606.10	Specialties
606.11	Equipment
606.12	Furnishings
606.13	Special Construction
606.14	Conveying System
606.15	Mechanical
606.16	Electrical
607	Complaints and Public Relations
608	Miscellaneous Correspondence
609	Schedules and Progress
609.1	Networks (often filed in separate chronological files for Networks and Printouts)
609.2	Printouts
609.3	Progress Reports (as required)
610	Insurance
611	Claims
611.1	Monetary
611.2	Delay
611.3	Undetermined, or Both Monetary and Delay
612	Accidents and Safety
613	Miscellaneous Construction Activities
614	Right of Entry(s)
614.1	Right of Way

614.2	Easements
614.3	Restoration
615	Surveys
616	Inspection
616.1	Inspectors' Daily Reports
616.2	Special Reports
617	Testing
618	Public Agencies
618.1	Building Authorities
618.2	Utility Companies
619-698	Expand as Required
699	AECM and Field Office Matters: Chrono Outgoing Chrono Incoming
699.1	General Instructions to Field Office
699.1.1	Resident Engineer's Manual
699.1.2	Field Engineer's Instructions
699.1.3	Inspectors' Instructions
699.1.4	Office Procedures (as required)
699.2	Policies and Requirements
699.3	Field Office Administration
699.3.1	Requisitions
699.3.2	Purchase Orders
699.3.3.	Field Personnel Actions and Overtime
699.3.4.	Expense Accounts
699.3.5	Accountable Property and Equipment
699.3.6	Construction Vehicles
699.4	Contracts
699.4.	AECM-Client Agreement
699.4.2	Subcontracts
699.4.3	Contracts
699.5	Manpower and Budgets
699.6	Miscellaneous Correspondence (Expand as Required)

Either File System A or B can be used effectively in AECM work. System A is preferred for smaller, less complicated projects, especially where the Resident Engineer has limited clerical staff. On complex projects, System B should be used. A competent secretary can be an invaluable aid to the Resident Engineer in filing and other work.

F. HANDLING CORRESPONDENCE

Even if the Resident Engineer has an extremely competent secretary, procedures for routine and filing correspondence shall be established or reviewed by the Resident Engineer and Project Manager to insure consistency of action.

To this end, it is desirable to have correspondence flow charts (as required) such as that shown in Figure 10.2.14. The Resident Engineer is responsible for such action.

VII. RECORDS AND REPORTS

A. CONTRACT DOCUMENTS

It is the responsibility of the Resident Engineer to insure that a complete and up-to-date set of plans and specifications is maintained in the field office at all times. Revisions of the original contract documents shall be posted and referenced, as appropriate.

At the start of each project (contract), the Resident Engineer will estimate and obtain the number of "working copies" of plans and specifications required by the field office.

Note that if the field office initially obtains one full-size reproducible of all contract drawings, plus one full-size set of drawings, together with any reproducibles or copies of reduced-size drawings which the Resident Engineer elects, it will be possible for the field office to reproduce any number of additional copies required.

B. CONTROL OF CONTRACT DOCUMENTS

1. Contract Drawings

Upon receipt of conformed drawings (copies or reproducibles) of the as-bid contract full-size drawings, these drawings (copies or reproducibles) will be logged and filed as record prints.

Upon receipt of revised drawings, the superseded drawings will be so marked, filed, and retained for record.

The receipt of all AECM team drawings will be logged in the "Drawing Control Log" promptly. (Figure 10.2.15.) Separate sections of the Drawing Control Log shall be established so that drawings which pertain to pending (not authorized) changes are held and logged separately.

2. Contract Details

Contract Details, so far as possible, shall be handled identically with Contract Drawings.

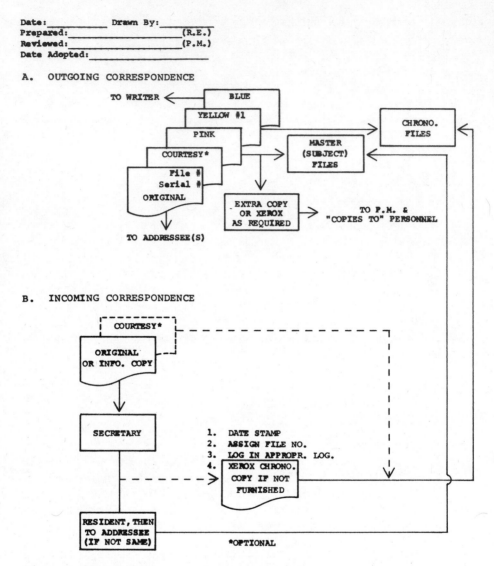

Figure 10.2.14. Sample correspondence flow chart.

DESIGN INC.

DRAWING CONTROL SHEET

Contract No.: _____
Title: _____

Page ____ of ____

DRAWING NUMBER	DRAWING TITLE		DRAWING REVISIONS							REMARKS
			REV. 0	REV. 1	REV. 3	REV. 4	REV. 5	REV. 6		
		NPC								
		DATE								
		NPC								
		DATE								
		NPC								
		DATE								
		NPC								
		DATE								
		NPC								
		DATE								
		NPC								
		DATE								

NPC = Notice of pending change or other identifying number.
DATE = Date drawing completed; ready to issue.

Figure 10.2.15. Sample of Drawing Control Log.

3. Contract Specifications

All conformed or as-bid sets of Contract Specifications shall be either furnished the Resident Engineer or prepared by the field office after proposals are received on each project (contract).

All addenda shall be posted and referenced in these specifications.

All subsequent changes to specifications shall be posted and referenced to documents making the change. Pending revisions to specifications shall not be posted until formally accepted.

C. PROCESSING CONTRACTOR(S) SUBMITTALS

"Submittals," as used herein, shall mean all show drawings, working drawings, samples, catalog data, certificates, field sketches or drawings, calculations, etc.

Usually, the specifications will enumerate all shop drawings required to be furnished by the contractor. Nevertheless, the Resident Engineer or Project Manager may have additional rights to require shop drawings under the "Materials and Workmanship" clauses of the Contract. The Resident Engineer is to be thoroughly familiar with the contract and these clauses, if applicable.

The Resident Engineer is responsible for receiving, processing, and expediting review of all shop drawings. These drawings shall be checked by the responsible designers; nevertheless, field office personnel shall check the submission in an appropriate fashion prior to transmittal through the Project Manager to the responsible designers. Shop drawings shall be thoroughly checked for conformance with all contract requirements; the Project Manager, Resident Engineer, and responsible designers shall insure this.

As soon as practicable after award of a contract, the Resident Engineer shall request the contractor to furnish a list of all shop drawings he plans to submit for approval, along with estimated dates of submission. (This may be part of the Project Network Schedule.) It is important that the list be complete and that shop drawings be submitted promptly on schedule. Timely and complete submittals will contribute materially to successful completion, and cooperation in this will be requested of the contractor. If the contractor seems to be dilatory or negligent in furnishing shop drawings, the Resident Engineer shall so advise him.

Typically, the contract will require the contractor to provide the AECM team, through the Resident Engineer, with a minimum of one reproducible and three copies (or seven catalog cuts) of all required shop drawings or working drawings. Additional reproducibles and/or copies may be requested of the contractor by the Resident Engineer for the purpose of obtaining or satisfying requirements of permits or authorities.

It is and will be the contractor's responsibility to make or provide all submittals and/or to obtain them from his suppliers and subcontractors.

The Resident Engineer will deal only with the prime contractor(s) unless he is requested or permitted to do otherwise by the contractor(s). The Resident Engineer shall endeavor to insure that submittals are made on time and that, once received, they are expeditiously handled by persons reviewing them to this end. The Resident Engineer shall establish and maintain a "Drawing Control Log-Submittal Section(s)" and shall establish a systematic suspense and follow-up procedure for handling submittals.

Varied systems can be developed for control of contractor submittals, and early establishment of procedures with the contractor is very desirable.

Depending on the anticipated number of submittals, the Resident Engineer should request the contractor to employ one of the following systems for transmitting his submittals:

1. A consecutive-numbered transmittal system.
2. A consecutive-numbered transmittal system with nonconsecutive or consecutive drawing numbers assigned.
3. A consecutive-numbered transmittal system with alpha-numeric consecutive or non-consecutive drawing numbers assigned.

These consecutive, non-consecutive, or alternatively alpha-numeric numbers are in addition to the suppliers' or vendors' control numbers, although they need not be if the contractor and his suppliers and subcontractors participate in this effort.

This coordinating action is left to the discretion of the Resident Engineer, but it is noted that, through cooperation with the contractor, the tedium of logging can be substantially reduced.

As an *example*, assume that the contractor has just transmitted a submittal of three shop drawings (one reproducible and two copies) to the AECM team. The following actions should be taken:

1. Upon receipt of the drawings (submittal), the Resident Engineer will insure that they are date stamped and logged in "as received." Use of a "Contractor Print Index" control log (see figure 10.2.16), is suggested.
2. Typically, the AECM team's contracts or subcontracts will require that one reproducible and from two to four prints be submitted. Receipt of this number is verified through logging. The Resident Engineer will establish a standard procedure to check the number of drawings received and to request additional drawings, if required.

CONTRACTOR PRINT INDEX

Contract No.: _____
Contractor: _____
Subcontractor: _____
Page ____ of ____

CONTR. DWG/SEQ NUMBER	DRAWING TITLE	QTY.	TYPE	CONTR. PRINT NO.	REV. NO.	RECEIVED	TRANS. NO.	SENT TO	REC'D FROM	SENT TO CONTR.	APPR. STATUS	REMARKS

TYPE: R = Reproducible
        P = Print
        C = Catalog Cut
        M = Eqpt. Manual
        S = Sample

APPR. STATUS: 1 = Approved
                2 = Approved as marked
                3 = Not approved
                4 = Partially approved

Figure 10.2.16. Sample of Contractor Print Index control log sheet.

3. Generally, the Project Manager will insure that the responsible designers review these submittals for conformance with contract requirements (design requirements, if need be). Nevertheless, the Resident Engineer will also review or assign responsibility for field review of drawings to members of his staff.
4. The Contractor Print Index must be completed in its entirety, and the submittals (drawings) marked similarly.
   a. Assignment of drawing sequence numbers is to be indicated in the *first* column of the form under the title, "Contractor Drawing Number," and this number is to be imprinted on each submittal.
   b. Under the fifth column, entitled "Contractor Print Number," the identifying numbers or marks used by the contractor or his subcontractor(s) should be repeated.
   Although it would be advantageous for the contractor and all subcontractors to use the drawing sequence numbers for print numbers, this generally is not feasible, especially with large contracts and large corporation suppliers or subcontractors who have pre-established print numbering systems.
   c. Completion of the "Drawing Title" column should match the submittal.

Completion of the rest of the Contractor Print Index is self-explanatory.

A simple visual display of appropriate filing of all drawings is given in Figure 10.2.17.

D. SUBMITTAL REVIEW AND APPROVAL

Upon receipt of submittals, the Project Manager shall insure that they are properly reviewed and approved or otherwise handled by the responsible designers. Additionally, the Project Manager or the Resident Engineer may, with authorization in certain special or emergency cases, assume responsibility for approval, depending on many factors including circumstances and the degree of technical significance.

Since AECM team responsibility for submittal review may vary through the contract and because of industry and professional practice, the entire AECM team shall give submittal review proper attention.

1. Objectives of Submittal and Review
   It must be realized that, generally:
   a. Contractor submittals indicate in detail how the contractor proposes to perform the various items of work.
   b. The submittal indicates the contractor's understanding of what is called for in the plans and specifications.

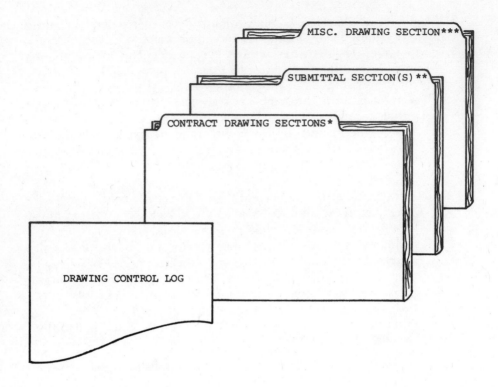

*   Contract Drawing Section - shall employ pre-printed drawing
    control sheets similar to Figure 10.2.15.

** Submittal Section - shall be organized and tabulated by
    Contractors and Subcontractors or by transmittal sequence
    or by specifications sections or other appropriate group-
    ings which may be designated by the Resident Engineer.

    Standard pre-printed Contractor print index sheets similar
    to Figure 10.2.16.

*** Miscellaneous Drawing Section - shall be used if applicable
    or if desired.

Figure 10.2.17. Sample drawings filing system.

    c. Approval of these drawings normally results in their becoming Contract Drawings.

    d. Direction to vary from these approved drawings becomes a Change.

        The importance of this approval and review work is apparent. The AECM team will, and must accept this important and proper responsibility.

        Nevertheless, in the interest of all parties involved, especially the AECM team, the Project Manager and Resident Engineer shall insure that the contractor is specifically advised as follows:

Approval of submittals shall not signify or be construed as a complete check, but will indicate only that the detailing and/or general methods of construction are satisfactory. Approval of submittals (shop drawings) will not relieve the contractor of the responsibility for furnishing material and equipment in compliance with the requirements of the contract plans and specifications or for any error which may exist on the submittals; the contractor shall be responsible for submittal dimensions and detailing and for the satisfactory construction of all work.

Further, if the contractor makes submittals which involve changes or performance in variance with contract plans and/or specification requirements, the variance must be called to the attention of the AECM team; otherwise, performance of the work as varied may not be acceptable.

2. Stamping of Submittals

    Generally, the responsible designers will review and stamp submittals to indicate that they are:

    a. Approved.

    b. Approved as Corrected.

    c. Partially Approved, as Noted.

    d. Disapproved.

    Rubber stamps indicating the approval status, as outlined above, are usually available or procured as required.

    Typically, words to the effect that: "Approval does not relieve the Contractor of the responsibility for the accuracy of the document or for full compliance with contract requirements," or a similar statement of equal intent, shall be included on this stamp.

3. Flexibility in Methods and Procedures

    AECM team members, when reviewing requests for approval of methods and procedures for accomplishing parts of the work by the contractor, shall allow a reasonable and practicable flexibility in such areas of work as:

      • Control of ground water.

      • Compaction of earthwork.

- Foundations.
- Underpinning.
- Structural framing.
- Reinforcing.
- Prestressing.
- Concreting.
- Installation of HVAC.
- Plumbing.
- Electrical.
- Vertical transportation.
- Similar types of work utilizing special equipment and/or materials.

AECM team members reviewing submittals in such areas shall use care not to restrict the contractor to use of requested methods and procedures *unless* conditions require restriction of this flexibility. Approvals of this type work and/or contractors' methods or procedures shall generally be made contingent upon the contractor obtaining the satisfactory specified result.

E. PROJECT RECORDS AND REPORTS

To keep track of day-to-day details of the work not covered by correspondence, and to permit recall of them after some time has elapsed, certain records and reports are to be taken and/or maintained in the project office. These records and reports are outlined and discussed in the following text.

1. Master Construction Diary (Resident's Daily Dairy)

The Master Construction Diary (Figures 10.2.18.A and B) should contain a record of all items of importance (even of routine matters covered by correspondence or other reports when circumstances are unusual), such as conferences with the contractor or other parties, agreements made, special notes regarding equipment or organization, labor conditions, weather, or other causes of possible delays, and any other matters that have a bearing on the history of the job. Instructions issued to the contractor and the contractor's reaction and pertinent remarks shall be entered with particular care.

Since this diary could have a strong bearing on contractor claims for extra compensation at job conclusion, or for denials of responsibility for such things as job operations, delays, or property damage, etc., to the extent feasible the diary must be well organized, with all entries being neatly made and clearly defined.

This diary shall be retained in the field office until project finalization; then it is to be transmitted to the Project Manager to become

MASTER CONSTRUCTION DIARY

PROJECT or CONTRACT NO.:_____ TITLE:_____

Hindrance to Normal Progress of Work: (*Discussions & Problems*)			TRADE	No. of Men	
				GEN. CON.	SUBS
			Supts/Assts		
			Foremen		
			Office		
Official Visitors:			Engineers		
			Carpenters		
			Laborers		
			Cem.Finish.		
General Notes, including Material & Labor Shortages:			Teamsters		
			Op.Eng/Oilers		
Work Progress - Subcontractors:		Compl.|Today	Steel - Reinf.		
Sub Name & Work	Description of Work	Qty.|Units	Steel - Struc.		
			Masons		
			Painters		
			Plaster/Lath		
			Plumbers		
			Heating		
			Electrical		
			Sht.Met/Roof.		
			Glass/Glazing		
			Tile Setters		
			Sub-Subs		
			(Show Names)		
Work Progress - Prime Contractor		Compl.|Today			
Items of Work	Description of Work	Qty.|Units			
			TOTALS:		
			WEATHER: Temp._____ Fine___ Rain___ Snow___		
			ACCIDENTS:_*_____		
			CHANGES:_____		
			TESTS:_____		
			PICTURES:_____		

Report Equipment on Reverse Side   *Give details in General Notes.

Report No._____

Date_____   Signed_____

Figure 10.2.18.A. Master Construction Diary (front page).

MASTER CONSTRUCTION DIARY
(continuation)

EQUIPMENT ON JOB

EQPT. NO.	DESCRIPTION	USED BY AND/OR FOR	* O/I	RATE PER (If Avail.)	HOURS USED	NOTES & COMMENTS

* O = Operated and maintained.
   I - Idle on jobsite.

Use balance of space to continue notes from reverse side, if required:

_____

_____

_____

_____

_____

_____

_____

_____

_____

_____

_____

_____

Figure 10.2.18.B. Master Construction Diary (reverse side).

part of the project permanent record. For multi-contract projects, more than one diary will be necessary. Handling of the diary with respect to security shall be consistent with the contents therein.

A suggested format for the Master Construction Diary is shown in Figures 10.2.18.A and B. The Resident Engineer may use this or an adaptation of this report format as he sees fit.

The intent of this report is to indicate in condensed form the nature and extent of any and all activity during the report period. Any unusual or abnormal circumstances must be described in detail. If necessary, supplemental sheets will be employed.

In general, separate report sheets should be made out for each prime contract or for each direct subcontract with the AECM team.

If more than one shift is worked at any time during the project a separate report shall be prepared for each shift's operation.

2. Inspector's Daily Report

An Inspector's Daily Report form shall be prepared by all inspection personnel for each contract or work item inspected. In cases of disputes with the contractor, structural failures, or even litigation, these reports can become important reference documents. These reports, therefore, must be filled out completely, as well as being grammatically correct and neatly prepared.

Figure 10.2.19 shows an illustration of an Inspector's Daily Report as it might be prepared following an inspection of the main building excavation support for our hypothetical Central City University New Campus Development project.

The Inspector's Daily Report form shall be identified by the name of the inspector, contract day or report number, job or contract title and/or number, week day, date, and shift, and should include information on:

a. Work performed that day (including type, amount, and labor count).
b. Type of weather and temperature if illustrative.
c. Equipment (both idle and employed).
d. Pertinent information on progress of work, delays, causes of delays, and extent of delays.
e. Instructions given to the contractor, including the name(s) of contractor's representative(s) to whom they were given.
f. Controversies, including pertinent details.
g. Safety violations observed and corrective measures taken; reports of any accidents or injuries.
h. Visitors to the project.

Constructioneering Northwest Inc.
Engineers

INSPECTOR'S DAILY REPORT

REPORT Nº _259_

DATE _July 7, 1976_

CONTRACT & Nº _12345_ _____ SHIFT _Day_ WEATHER _Fair_ TEMP _60°_

CONTRACTOR _ABC Company_ _____ CONTRACTOR'S REPR. _Sharp_

FEATURE(S) _Support of Excavation - Main Building_

PROGRESS OR DESCRIPTION OF WORK:

1 Clam, 1 Dozer and 1 F.E. Loader excavating north corner.
The F.E. loader was broken down for 2 hours.
Lagging continues throughout work areas.
Welding continues in all areas and is getting further behind every day.
General cleanup of all working areas was carried out.

SPECIAL INSTRUCTIONS FROM SUPERVISOR:

Check for over-excavation on north end.

UNUSUAL OR UNSATISFACTORY CONDITIONS:

Bottom at and around northeast corner lagging; getting boggy and wet.

DELAYS (length & reason):

2 hr. delay, F.E. loader broken down.

REQUESTS FROM CONTRACTOR:

None.

INSTUCTIONS TO CONTRACTOR:

Advised A.B. Sharp not to excavate any deeper on northeast end until
struts are installed. Excavation continued at north end, however.

CONTRACTORS FORCES & EQUIPMENT:

		ABC Manual	Non-Manual	Total
1 Clam	1 Compressor			
2 Arc Welders	2 Cranes			
1 Dozer	12 10-wheel dump	62	4	66
1 F.E. Loader	trucks			

REMARKS (needs, accidents, visitors, etc.):

B. Smith injured when struck on left shoulder by a 2" x 12" x 3' board
which fell from street level to bottom of excavation. Time approximately
4:20pm. Possible lost time accident. Smith driven to Doctor's Clinic
by Safety Engineer Brown.

SIGNATURE _John Blue_

Figure 10.2.19. Sample Inspector's Daily Report.

The Resident Engineer is responsible for the completeness, neatness, and accuracy of Inspector's Daily Reports. He shall, therefore, carefully review all reports to insure that they are completed properly.

These reports shall be filed and retained in the field office until project (or contract, if applicable) completion, at which time they shall be forwarded to the Project Manager for incorporation into the project permanent record.

3. Weekly Progress Report

The Resident Engineer will report the progress of the project(s) under his control weekly on a Weekly Progress Report form (see Figures 10.2.20.A and B).

The reporting period shall begin on Monday and end on Sunday, and the report is to be submitted to the Project Manager no later than Monday afternoon following the Sunday ending the week being reported. Specific items to be covered in each report should include, but are not limited to, the following subjects:

   a. Report number and reporting period covered.
   b. Scheduled and / or actual construction status.
   c. Weather conditions.
   d. Manpower summaries; weekly average personnel (AECM team field employees, as well as contractor(s) and subcontractor(s) personnel).
   e. Construction equipment.
   f. Construction status and work in progress narrative reporting.
   g. Visitors.
   h. Current or potential job problems.
   i. Job photographs.
   j. Miscellaneous, including reports, graphs, and charts, as appropriate.

Items i and j listed above may be deferred from the Weekly Progress Report and furnished monthly with the concurrence of the Project Manager. The details of these Weekly Progress Reports are to be worked out between the Resident Engineer and the Project Manager.

4. Monthly Progress Report

The need for a Monthly Progress Report may be established by the Project Manager. If required, it shall be a comprehensive narrative which could include photos, diagrams, charts and/or graphs which would graphically represent the work completed during the period being reported and that completed to date. The intent of these

WEEKLY PROGRESS REPORT

WEEKLY PROGRESS REPORT NO. ___10___ FOR THE WEEK OF: Oct. 15-21, 1978

CONTRACT NO. __12345__ CONTRACTOR: __ABC Company__

CONSTRUCTION OF: __University - Library Building__

SCHEDULE STATUS:

     Status to Date: Overall project 2 wks. behind original CPM schedule.
     Scheduled Completion: October 15, 1978
     Anticipated Completion: November 1, 1978

PERCENT COMPLETED:

     To Date: 53%
     Scheduled: 54%

WEATHER:

     Clear and mild (70° to 45°), except for light rain on Wednesday; no
     closing down work. Wet grounds on Thursday slowed earthwork opera-
     tions during the morning only.

PERSONNEL:	Manual	Non-Manual	Total
ABC Company - Contractor	192	15	207
Brown Company - Grading Sub	43	3	46
Smith Company - Struct. Erect. Sub	5	1	6
	240	19	259

Note: Work performed on a 5-day/8-hour basis.

CONSTRUCTION STATUS:

     Library Building Foundation

         On Wednesday, October 17, Brown Company, grading subcontractor to
         ABC Company, started excavation work.

     Foundation Concrete Work

         Continued forming north wall 1st and 2nd lifts (132 c.y. placed this
         week). Smith Company, structural erection subcontractor to ABC Com-
         pany, set 4 base plates for exterior columns on Thursday, October 18.

Figure 10.2.20.A. Sample Weekly Progress Report (front page).

WEEKLY PROGRESS REPORT - PAGE TWO

PROJECT VISITORS:

    L. M. Morris, General Manager, XYZ Cement Co. - October 16
    A. M. Finn, Engineer, Municipal Building Department - October 17

CURRENT OR POTENTIAL JOB PROBLEMS:

    ABC Company is experiencing difficulty in obtaining qualified welders;
however, no schedule delay has occurred to date.

REMARKS:

SIGNATURE: A. R. Best
                   Resident Engineer

DATE: October 22, 1978

Figure 10.2.20.B. Sample Weekly Progress Report (reverse side).

graphic illustrations and/or photographs is to permit visual understanding of work performed.

A complete descriptive illustration of a Monthly Progress Report for the C. C. University New Campus Development Project is furnished in Figures 10.2.21.A through 10.2.21.C. Then, in Figures 10.2.22 through 10.2.26, summary forms are presented which furnish supporting data to the Monthly Progress Report. These are illustrative of the type information which may be appended to the reports.

If a need is established for such monthly progress reporting, it is the responsibility of the Resident Engineer to see that the report is submitted to the Project Manager in a timely manner, and that the information contained therein is presented in a neat and orderly fashion, with all pertinent data accurately detailed and documented, where appropriate.

5. Test Reports
   The Resident Engineer will insure that complete and accurate reports of all tests performed are made, and that a file of these reports is maintained in the project office. These tests and Test Reports, whether made by the AECM team field personnel or by an outside subcontractor, must be handled professionally, since they may be required in courts of law. Test Reports should include, but may not necessarily be limited to, the following items:

- Contract title and number.
- Description of contract work item tested.
- Name of the person and the company performing the test, location, date, and equipment used.
- Sample source and date secured.
- Narrative description of tests performed.
- Results of test(s).
- Recommendation as to acceptance or rejection.
- Signature of responsible person performing or controlling test work.

A sample Test Report form is provided in Figure 10.2.27.

6. Field Notes
   a. Field Notes shall be made out for:
      (1) Field Surveying (except contractor's surveying).
      (2) General Project Record (as applicable; similar to a diary).
   b. If practical, separate Field Notebooks should be used for the following activities:
      (1) Original Conditions Surveys and Observations.
      (2) Layout Work.

MONTHLY PROGRESS REPORT

MONTHLY PROGRESS REPORT NO. 2   FOR MONTH ENDING: May 30, 1978

AECM FIRM   JOB NO. 1234 CONTRACT: Univ.   New Campus Facilities

A. NARRATIVE OF PROGRESS

    1. Summary of Architectural & Engineering Work

        a. The Architectural Concept and Design Group is preparing draw-
           ings showing possible Library Office arrangements. The land-
           scaping and paving layout at the north entrance of the Admini-
           stration Building is being studied.

        b. The Design Drawing Group is preparing final drawings for pre-
           cast concrete work. Bid drawings for elevators, steel stairs,
           and acoustical ceilings were completed this month.

        c. HVAC, electrical, and plumbing designers are continuing design
           development work.

    2. Current Status of Construction

        a. DESCRIPTION OF WORK PERFORMED SINCE LAST PERIOD

           (1) Permanent job office facilities were set up by AECM team
               and prime grading contractor.

           (2) Barricade system and fencing enclosing entire project site
               was completed.

           (3) Temporary electrical feeder line installation was completed
               to the transformer pad location.

           (4) Pile driving equipment and temporary office trailer was
               moved in. Eight (8) test piles were driven May 15.

           (5) Survey control and base line layout completed. Four (4)
               permanent control monuments set.

           (6) Clearing and grubbing of the south end of site was started
               May 2.

           (7) Excavation and lagging for Library Building basement started
               May 3, and is 60% complete.

        b. CONTRACTS NEGOTIATED & AWARDED DURING PERIOD

           None.

        c. CONTRACTORS, SUBCONTRACTORS & MAJOR SUPPLIERS

           (1) Pacific Map, Inc. - Land Surveying

Figure 10.2.21.A. Sample of Monthly Progress Report for University
Project (first page).

MONTHLY PROGRESS REPORT - PAGE TWO

    (2)   ABC Construction Company - Prime Sitework & Foundations

        (a)   Brown Company - Grading Subcontractor

        (b)   B.M. Long Company - Piling Subcontractor

        (c)   Smith Company - Structural Erection Subcontractor

    (3)   Red Iron, Inc. - Piling Supplier

  d. CONSTRUCTION SCHEDULE, CHANGES & EXTRAS

    (1)   ABC Construction submitted a bar chart for initial work (60 workdays), and is preparing a network plan at this time.

    (2)   Mobilization, clearing, and excavation work are just starting up and seem to be on schedule.

    (3)   No changes or extras to be noted this period.

  e. GENERAL WORK CONDITIONS AND REMARKS

    Construction just starting. Initial work seems to be going along as well as could be expected.

B. JOB PHOTOGRAPHS

None processed as yet; will be included in next monthly report.

C. REPORTS AND GRAPHS

Since work has just started and graphic trends cannot be indicated adequately at this early stage, this section of this monthly report will summarize what will be provided in the future and will, in abbreviated forms, represent current statistics.

Reports and/or graphs for the following areas should be included:

1. Production - Work Accomplished

Production in general will be measured in what is determined to be an appropriate measure by the Resident Engineer. Graphs of actual versus scheduled production of key items shall be presented, such as:

a. Evcavation.

b. Concrete poured.

c. Steel set, etc.

For production graphs for this period, see Figures 10.2.22 and 10.2.23 which follow.

Fig. 10.2.21.B. Continuation of Sample Monthly Progress Report for Central City University Project (page two).

MONTHLY PROGRESS REPORT - PAGE THREE

2. Construction Costs & Cost Trends

Construction costs and cost trends reporting procedures are being
finalized with University. At this time, it is envisioned that
this portion of the monthly report will be a summarization of the
current and forecasted status of the project in terms of both
quantities and costs. A cash flow report is being developed and
will be included in next month's report.

Three complete cost report forms are attached. These reports are
prepared in sample, or draft format for the following:

a. COST SUMMARY

    This form summarizes the other cost reports and shows the
    project's current and forecasted costs. (See Figure 10.2.24.)

b. CONSTRUCTION & MATERIALS SUPPLY CONTRACT SUMMARY

    This form summarizes and shows the status of awarded and future
    construction and materials supply contracts. (See Figure 10.2.25.)

c. CONTRACT COST SUMMARY

    This form shows current costs and forecasts by contract or,
    preferably, by contract bid item. This form is extremely adapt-
    able to unit price contracts but, in preparation of an accurate
    forecast, a diligent effort to evaluate quantities to complete
    will have to be made. (See Figure 10.2.26.)

    Quantities must be determined accurately through observation of
    field conditions, spot checks of drawings and other appropriate
    calculations.

    The current and forecasted costs of all existing or anticipated
    extras and change orders shall be shown. A brief description or
    explanation shall be given for each anticipated extra work order
    or change order.

3. Resource Commitments

    This section shall include a statement of the commitments of all
    resources -- men, materials, and equipment.

Fig. 10.2.21.C. Continuation of Sample Monthly Progress Report for
Central City University Project (page three).

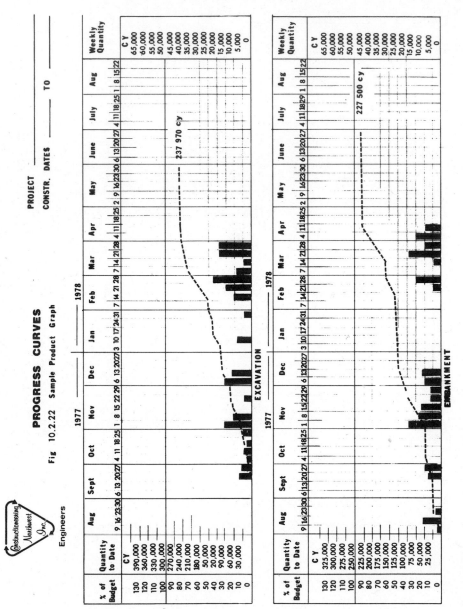

Figure 10.2.22 Sample Product Graph

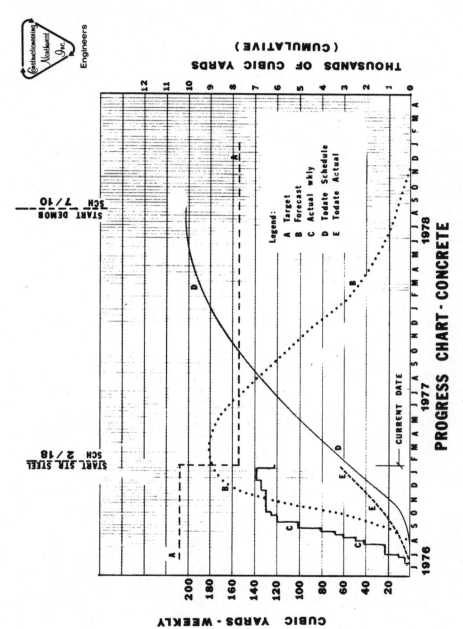

Figure 10.2.23. Sample Production Graph in Support of Other Progress Reports.

COST SUMMARY AS OF _____

Job No.:   Project: Client:

DESCRIPTION	PREVIOUS COSTS	COSTS THIS PERIOD	COSTS TO DATE	FORECAST OF TOTAL COSTS	REMARKS
Contracts Awarded	$	$	$	$	
Contracts Future					
TOTAL CONTRACT COSTS	$	$	$	$	
AECM Costs					
Architecture					
Engineering					
Construction Management					
Field Constr./Management					
TOTAL AECM COSTS	$	$	$	$	
Client Costs					
Land & Right of Way					
Wages					
Interest on Money					
Other					
TOTAL CLIENT COSTS	$	$	$	$	
GRAND TOTAL COSTS:	$	$	$	$	

Figure 10.2.24. Sample of Cost Summary form to accompany the Monthly Progress Report.

CONSTRUCTION & MATERIALS SUPPLY CONTRACT SUMMARY AS OF _____

Job No.: _____ Project: _____ Client: _____

DOCU-MENT NO.	CONTRACT DESCRIPTION	CONTRACTOR	CONTRACT BID AMOUNT	CHANGE ORDER AMOUNT	BID AMOUNT + C.O.'s	COSTS THIS MONTH	COSTS TO DATE	FORE-CAST COSTS	NOTICE TO PROCEED	COMPLETION DATE ORIG-INAL	COMPLETION DATE EXTEN-SION	COMPLETION DATE REVISED
	CONSTR. CONTRACTS AWARDED:											
	SUB-TOTALS											
	MATERIALS & SUPPLY CONTRACTS AWARDED											
	SUB-TOTALS											
	TOTALS AWARDED:											
	CONSTR. CONTRACTS FUTURE:											
	SUB-TOTALS											
	MATERIALS & SUPPLY CONTRACTS FUTURE											
	SUB-TOTALS											
	TOTALS FUTURE:											
	GRAND TOTALS:											

Figure 10.2.25. Sample of Construction & Materials Supply Contract Summary form to accompany the Monthly Progress Report.

CONTRACT COST SUMMARY AS OF _____

Job No.:
Project:
Client:

Contract No.:
Contractor Name:
Contract Work:

Document No.: _____
Page _____ of _____

BID ITEM NO.	DESCRIPTION	UNIT	QUANTITY				CONTRACT PRICE	COSTS						
			THIS PERIOD	TO DATE	FORECAST TO COMPL.	F'CAST TOTAL	BID	THIS PERIOD	TO DATE	FORECAST TO COMPL.	F'CAST TOTAL	BID	DIFF. FROM PREV. F'CAST	
SUB-TOTALS:														
GRAND TOTALS:														

Figure 10.2.26. Sample of Contract Cost Summary

Field Data On Concrete Compression Test Specimens

Date Moulded: _____ , 19 ____

Project: _____

At: _____

Contractor: _____

Reported to: _____

Concrete Supplier: _____

Quantity represented _____ Cu. Yds. Specimens made by: _____ No. Submitted: _____

Location of pour _____

Strength requirement _____ psi at 28 days

MATERIAL PROPORTIONS USED (Quantities per cubic yard of concrete) MIX No.: _____

Cement _____ lbs. _____ bags Type _____ Brand _____

Fine Aggregate (SSD) _____ lbs. Source _____ Type: _____

Coarse Aggregate (SSD) _____ lbs. Size _____ Source _____ Type: _____

Coarse Aggregate (SSD) _____ lbs. Size _____ Source _____ Type: _____

Water Total _____ Gals. _____ Gals./bag

Admixture: Amount _____ Kind _____

Admixture: Amount _____ Kind _____

Above information obtained from: batching inspection ☐   Project Engineer ☐   Contractor ☐

Concrete Supplier ☐   Other ☐ (Designate) _____

Entr. Air _____ %   Actual Slump _____ In.   Ticket No.: _____

Temperatures: Air _____ °F Concrete_____ °F Weather _____

Mixing: Central Mix ☐   Truck Mix ☐   Job Mix ☐

Above Site Information obtained from: Inspection ☐   Project Engineer ☐   Contractor ☐

Concrete Supplier ☐   Other ☐ (Designate) _____

Cyls. delivered to laboratory by: CNI ☐   Contractor ☐   Engineer ☐   Common Carrier ☐

SPECIMEN MARKING	TEST AT DAYS	LAB NO.	DATE REC'D.	DATE TESTED	TOTAL LOAD=LBS.	UNIT LOAD=P.S.I.	REPORT NO.
_____ A							
_____ B							
_____ C							
_____ D							
_____ E							
_____ F							

If test results are low, note below if there is any discernible cause.

Remarks: _____

Signed: _____

Figure 10.2.27. Sample Test Report form.

(3) Construction Progress Documentation.

(4) As-Built Recording.

Generally, these notes shall be kept neatly in booklet form, with the first few pages of the book reserved (left blank) for subsequent indexing. Procedures similar to those in general professional practice shall be followed.

The Resident Engineer will furnish standard prenumbered Field Notebooks to all members of his field staff and, if requested, to other AECM team members visiting the site.

All AECM team field personnel are to be encouraged to keep daily notes of their actions and observations. In addition, all AECM team surveying work, whether performed by AECM team personnel or by outside surveyors, shall be documented in these booklets, which shall be filed in the field office.

A sample sheet of field survey notes is shown in Figure 10.2.28.

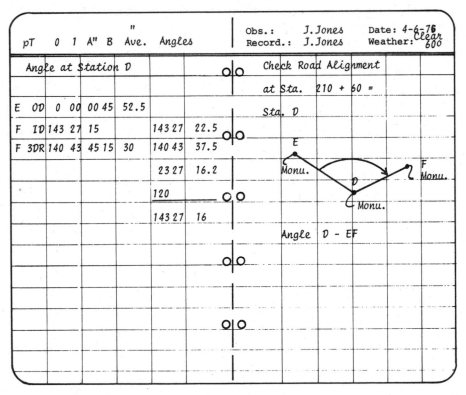

Figure 10.2.28. Sample Field Notebook pages with field surveying notes filled in.

7. Visitors' Register

The Resident Engineer and his staff shall extend all possible courtesy to and attempt to cooperate with all visitors. A Visitors' Register logbook shall be made available in the AECM team's field office, and all visitors shall be requested to sign it for the record. A standard, preprinted Visitors' Register form is available for this purpose, and a sample page is shown in Figure 10.2.29.

8. Progress Photos

The Resident Engineer and his staff shall take and file appropriate progress photos of the project. Generally, the photographic work will be augmented by aerial photographs taken on a monthly basis and maintained in the Progress Photo files or in a Photo Log for ease of reference.

Either a Photo Log or Photo File, or both, will be maintained in the AECM team's field office. The Resident Engineer shall see that a Polaroid-type or other suitable and acceptable type camera is available in the field office at all times.

All photos taken shall be marked to indicate the subject, date and the name of the person who took the photo. All photos taken by AECM team personnel or by an outside photographer (as in the case of monthly aerial photos) shall be filed or recorded in an appropriate log, as the Resident Engineer directs.

9. Other Documents and Forms

Project field services by the AECM team can often be facilitated through the use of form letters and standardized forms. The Project Manager and Resident Engineer shall work together in developing appropriate forms and control documents for the project.

In the several Figures which follow this portion of the text are illustrations of numerous forms which have proved of value in past work.

These forms are furnished for reference and/or for adoption in the field, as the Resident Engineer may deem it appropriate.

Accompanying the illustration of each of the following forms are instructive notes in the use of the forms, giving, in some cases, detailed information of preparation of the forms. Other illustrations may be accompanied only by brief notes explaining the purpose or intent of the form; in these cases, completion of the form is self-explanatory.

Visitors' Register

DATE	NAME	FIRM	ADDRESS	TO SEE
			Street	
			City State	
			Street	
			City State	
			Street	
			City State	
			Street	
			City State	
			Street	
			City State	
			Street	
			City State	
			Street	
			City State	
			Street	
			City State	
			Street	
			City State	
			Street	
			City State	
			Street	
			City State	
			Street	
			City State	
			Street	
			City State	

Figure 10.2.29. Sample page from Visitors' Register logbook.

PERSONAL DAILY DIARY

The Resident Engineer and all his personnel are encouraged to maintain personal diaries to supplement other job records and reports.

190

Simplified Guide to Construction Management

Engineers

400 Building, 400 108th Avenue N.E. - Bellevue, Washington 98004
Phone (206) 455-9230 or 232-1190

PERSONAL DAILY DIARY

Weather: _Clear and Mild_   Report Day No.: _____22_____

Temperature Range: _45° - 70°_   Day & Date: _Monday, 10-13-76_

Ten-hour workday for ABC Company: 8:00a.m. - 6:00p.m.

Started excavation work on Main Building foundations. Continued concrete work. Completed curbing work on access to parking area, and resumption of vehicular traffic and parking anticipated next week.

Met with A.B. Sharp, ABC Company, at 10:00a.m. in AECM field office to discuss delay in starting Main Building foundations. He proposes to work 14 hours a day and Saturdays until caught up in approximately three weeks. He plans to claim for extras and wants to revise the critical path schedule to indicate this delay.

From 11:05a.m. to 12:15p.m., Brown Company, grading subcontractor for ABC Company, was moving in equipment.

Telephoned F. G. Hayes, of Hayes Office Interiors, at 1:30p.m. regarding the delivery of additional office furniture on order, and found they will be delayed 20 days. No problem; this is sufficient time for personnel being added in November and December.

Mr. Haley, of XYZ Forms, called our office at 3:20p.m. to report that balance of steel forms needed for next week's pours will be on jobsite Wednesday afternoon, October 15.

Telephoned Project Manager at 4:30a.m. to review job personnel requirements (one man promised November 1 and two men on December 1). Also, received his instructions to negotiate a change with ABC Company for delays on October 5.

Figure 10.2.30. Sample page from Resident Engineer's Personal Daily Diary as it might appear on the University New Campus Development Project.

SUMMATION OF WEIGHED MATERIAL TO DATE

Form 0210-S will be used to provide a cumulative total of weighed materials received on the project.

Quantities will be entered on this form from Form No. 0210, Daily Summary of Weighed Materials.

When completed, the Summation of Weighed Material to Date form may be used in verifying the contractor's quantities of materials paid for by weight.

The form will be prepared in duplicate in the Field Office. The original and one copy will be furnished to the Resident Engineer, who will forward one (1) copy to the Project Manager.

Constructioneering Northwest Inc.

SUMMATION OF WEIGHED MATERIAL TO DATE

Material _____ Report No. _____

Contractor _____ Page _____ of _____

Supplier _____ Contract No. _____

DATE 19	DAILY SUMMARY SHEET NO.	TONS	TOTAL TONS TO DATE	MATERIAL PLACED IN	REMARKS
	Total Brought Forward				
	Total Carried Forward				

Form No. 0210-S

Figure 10.2.31. Sample form for Summation of Weighed Material to Date.

DAILY SUMMARY OF WEIGHED MATERIAL

Form No. 0210 will be prepared by the Inspector upon receiving materials on the project for which payment under the contract is to be made by weight.

Weight tickets must be turned over to the Inspector and attached to the original of this form.

This form will be prepared in quadruplicate. One (1) copy is to be transmitted to the contractor and two (2) copies to the Resident Engineer, one of which will be forwarded to the Project Manager. The original will remain with the Resident Engineer.

If appropriate, this form may be filled in at the scale house, or the Resident Engineer may authorize an equivalent records system.

MATERIAL			MATERIAL FROM			PLACED IN		
Weight Ticket #	Truck No.	Net Weight of Material	Weight Ticket #	Truck No.	Net Weight of Material	Weight Ticket #	Truck No.	Net Weight of Material
							Total Lbs.	
							Tons Col. 3	
							Tons Col. 2	
Total Lbs.			Total Lbs.			Tons Col. 1		
Tons Col. 1			Tons Col. 2			Tons This Sheet		

DAILY SUMMARY OF WEIGHED MATERIAL

Material_____ Report No._____
Contractor_____ Page ____ of ____
Supplier_____ Contract No._____

Constructioneering Northwest Inc.

CORRECT _____ (Date)   WEIGHMAN _____ (Date)

Form No. 0210

Figure 10.2.32. Sample Daily Summary of Weighed Material form.

DAILY TRUCK TARE WEIGHTS

Form No. 0211 will be completed by weighman or scale master. The form will be prepared to indicate the tare weight of *all* trucks used in hauling materials to the project which are to be paid for by weight.

Tare weight must be obtained from a tested and sealed scale when measurement of an item is specified in "Measurement and Payment" to be by weight.

Tare weight for each truck used will be obtained once during each 8-hour period it is in operation, after major repairs, following alterations, or at such times determined necessary by the Resident Engineer.

The form will be completed in duplicate. The original will be turned over to the Resident Engineer daily. The other copy is to be retained by the contractor.

Constructioneering Northwest Inc.

DAILY TRUCK TARE WEIGHTS

Contract No._____   Location_____

Contractor_____   Date_____

Weighman_____   Weather_____

Weight Tolerance_____   Scale Capacity_____

NO.	TRUCK OWNER	REGISTERED GROSS CAPACITY	PREVIOUS DAY'S TARE	TODAY'S TARE		
				1	2	3

Form No. 0211

Figure 10.2.33. Sample Daily Truck Tare Weights.

WEEKLY SUMMARY OF FIELD DENSITY TESTS

Form No. 0215-S is a recapitulation of "Density Tests" taken daily or periodically during the week.

The form will be used as a permanent file record of densities obtained from compaction at various locations such as fills, backfills, grading, etc. It will be compiled from the Daily Field Density Test Report, Form No. 0215.

This form will be prepared in triplicate. The original and two (2) copies will be furnished to the Resident Engineer, who will forward one (1) copy to the Project Manager.

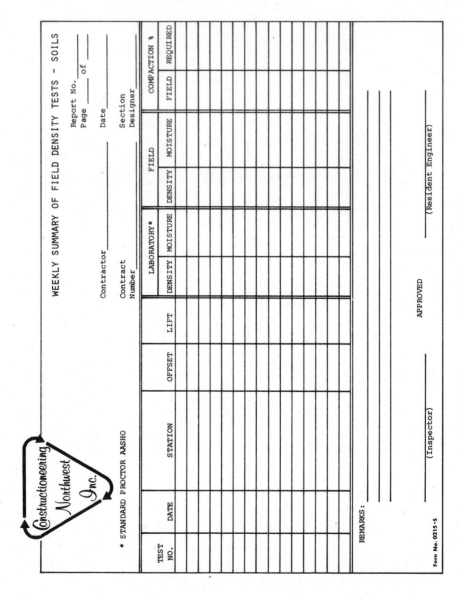

Figure 10.2.34. Sample Weekly Summary of Field Density Tests.

REPORT OF DAILY FIELD DENSITY TESTS

Form No. 0215 is to be used as the field office daily summary of density tests.

The figures on this form will be used in the compilation of the Weekly Summary of Field Density Tests, Form No. 0215-S.

This daily report form will be prepared in duplicate for the Resident Engineer, who will retain the original in his files and forward one (1) copy to the Project Manager.

Test results will be provided to the contractor as soon as available, on an appropriate form.

In work where two or more lifts of material are being placed, tests shall be conducted on each lift. In general, initial lifts in any one area should be acceptable prior to placement of subsequent lifts.

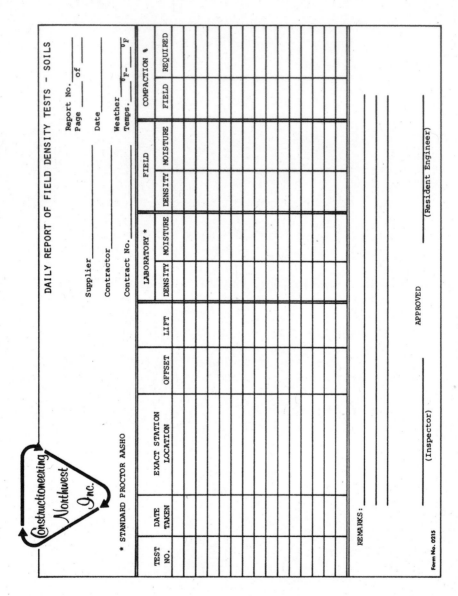

Figure 10.2.35. Sample Report of Daily Field Density Tests.

REPORT OF FIELD SIEVE ANALYSIS

Form No. 0220 will be prepared by the Inspector whose duty it is to make such tests. The form must be approved by the Resident Engineer.

This form may be used for sieve analysis of coarse or fine aggregate for either concrete or bituminous mixtures and for specified sub-base, base, or backfill materials.

The inspector preparing the form will insert, in the spaces provided, the sieve sizes used in the testing.

The form will be prepared in duplicate. The Resident Engineer shall retain the original for his files and transmit one (1) copy to the Project Manager.

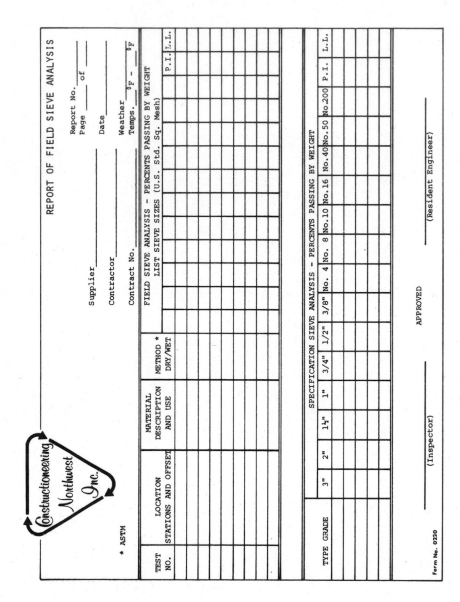

Figure 10.2.36. Sample Report of Field Sieve Analysis.

RECORD OF PILE LOADING TEST

This form shall be used to record the actions and results of load tests on a single pile or groups of piles.

Tests will be conducted in accordance with requirements of Special Conditions or applicable procedures from ASTM or local Building Authority.

Form No. 0230-PT will be prepared by the Field Inspector in duplicate. The record is to be checked by the Resident Engineer, who will retain the original for his files and forward one (1) copy to the Project Manager.

INSTRUCTIONS FOR MAKING PILE TESTS

Pile Tests—When called for or required, pile tests will be carried out as follows. Test pile(s) shall be driven and records kept of driving.

Damaged Piles—Damaged pile heads shall be cut back to sound material before the test is made.

Curing Cast-in-Place Piles—Unless otherwise provided, shells or tubes which, in service, are to be filled with concrete shall be filled with concrete in sufficient time for adequate curing before the test is made. High-early-strength cement may be used.

Application of Loads—Following completion of driving by not less than 48 hours, the test load shall be applied concentrically as near the ground as practicable, by direct weight or by hydraulic jack pressure that is automatically maintained.

If anchor piles or devices are used to provide reaction, the axis of the nearest such pile or device shall be located at least five (5) feet from the loaded pile. The initial and subsequent increments shall be applied gradually. The test load and its application shall be so arranged that readings can be taken

by the engineers directly on the pile and a record made of
the load and settlement simultaneously at any time.

Amount of Load—The piles to be tested shall be loaded to at
least twice the proposed working load. The initial load shall
not exceed two-thirds and the increments shall not exceed
one-third of the proposed or anticipated working load.

Records—Records shall be made of the load, time, and
settlement, as accurately as practicable, immediately before
and following a change of load. While noticeable settlement
is taking place, reading shall be at 10-minute intervals or
less; otherwise at hourly intervals. Use Form No. 0230-PT,
Record of Pile Loading Test, to report test information.

Loading Test—In making these tests, at least one (1) hour
shall elapse between the applications of successive increments. When the total load equals twice the proposed working load, it shall be left in place at least 48 hours.

Rebound—In determining net settlement of a tested pile,
deduction shall be made for the allowable elastic recovery of
the pile as indicated by the rebound within 12 hours following the release of the test load.

Constructioneering Northwest Inc.

RECORD OF PILE LOADING TEST

Report No. _____
Page _____ of _____

Foundation Designer _____
Contractor _____
Contract No. _____

Date _____
Shift 1 2 3
Weather _____
Temps. _____ °F - _____ °F

TYPE PILE _____
LOCATION OF PILE _____
DESIGN LOAD _____

LENGTH OF PILE _____
IN STRUCTURE YES _____ NO _____
200% DESIGN LOAD _____

TYPE OF HAMMER USED IN DRIVING _____
LENGTH PREBORED _____
LOADING MATERIAL _____

PILE NO.	LOAD LBS.	TIME AFTER LOAD APPLICATION/RELEASE TO OBSERVATION	CHANGE	
			−	+

PILE NO.	LOAD LBS.	TIME AFTER LOAD APPLICATION/RELEASE TO OBSERVATION	CHANGE	
			−	+

REMARKS:

_____ (Inspector)

APPROVED

_____ (Resident Engineer)

Form No. 0230-PT

Figure 10.2.37. Sample Record of Pile Loading Test.

RECAPITULATION OF PILES DRIVEN

Form No. 0230-S will be prepared to cover load bearing pile driving in any location, or, if appropriate, all piles in any one footing.

The form will be prepared by the Inspector or individual in charge of the pile driving operation and checked by the Resident Engineer.

This form will be completed in triplicate. The original will be retained by the Resident Engineer for his files, and two (2) copies forwarded by the Resident Engineer to the Project Manager.

RECAPITULATION OF PILES DRIVEN

Constructioneering Northwest Inc.

Shift ___1 ___2 ___3   Report No._____

Weather_____   Date_____

Contractor_____   Contract No._____

DRAWING NO._____   TYPE OF PILE_____ HAMMER_____

LOCATION_____   SPECIFICATION FORMULA_____

DATE DRIVEN	PILE NO.	BATTER	LENGTH IN FEET		ELEVATION		CALCULATED BEARING TONS
			IN LEADS	IN PLACE	CUTOFF	TIP	

NOTE: Show make-up of piles (i.e., 40+20+10).
Use back-up sheet for explanatory
sketches if necessary.

_____ (Inspector)

_____ (Resident Engineer)

Form No. 0230-5

Figure 10.2.38. Sample Recapitulation of Piles Driven.

PILE DRIVING RECORD

Two forms are provided here for use as a Pile Driving Record—Form No. 0230-A and Form No. 0230-B. One of these forms shall be used by the field inspection forces to keep driving records of each driven pile. The Resident Engineer shall select the form to be employed.

Information from the Pile Driving Record form shall be used in the preparation of Form No. 0230-S, Recapitulation of Piles Driven.

This form shall be prepared in duplicate. One (1) copy will be retained by the Resident Engineer and one (1) copy will be sent to the Project Manager.

Constructioneering Northwest Inc.

PILE DRIVING RECORD

Report No._____

Section _____   Page _____ of _____

Designer_____   Date_____

Contractor_____   Contract No._____

DWG. NO._____   LOCATION_____   PILE NO._____

TYPE OF PILE_____   MAKE & MODEL_____   RATED ENERGY_____

SPECIFICATION BEARING_____TON   SPECIFICATION FORMULA_____

BATTERED_____ VERTICAL_____ PIER_____ BRIDGE_____ ABUTMENT_____ WALL_____

OTHER_____   LENGTH PREBORED_____

Ground Elevation_____   Length in Leads_____

Tip Elevation_____   Penetration into Ground_____

Cut-Off Elevation_____   Length in Place_____

DEPTH BELOW GROUND	BLOWS PER FOOT	REMARKS	DEPTH BELOW GROUND	BLOWS PER FOOT	REMARKS
0-1			0-1		
1-2			1-2		
2-3			2-3		
3-4			3-4		
4-5			4-5		
5-6			5-6		
6-7			6-7		
7-8			7-8		
8-9			8-9		
9-0			9-0		
0-1			0-1		
1-2			1-2		
2-3			2-3		
3-4			3-4		
4-5			4-5		
5-6			5-6		
6-7			6-7		
7-8			7-8		
8-9			8-9		
9-0			9-0		

REMARKS :_____

_____

_____   APPROVED   _____

(Inspector)   (Resident Engineer)

Form No. 0230-A

Figure 10.2.39. Sample Pile Driving Record.

PILE DRIVING RECORD

This form, No. 0230-B, may be used as an alternate for Form No. 0230-A. The Resident Engineer shall determine which form should be used for recording pile driving information. See notes on page 000 pertaining to the use of this form.

Constructioneering Northwest Inc.

PILE DRIVING RECORD

Pile No._____

Date_____

Client	Project	Contractor	Job No.

	Pile		Hammer – Make & Model

Pile

Type	Hammer – Make & Model

Tip Diameter _____ in.	Butt Diameter _____ in.	Weight _____ lb.	Description
Length Driven _____ ft. _____ in.	Weight _____ lb.		

Dr.Cap, ttr,etc

Mandrel

Description	Started Driving ____ am ____ pm	Finished Driving ____ am ____ pm
Length _____ ft. _____ in.	Weight _____ lb.	Driving Time _____ min.

Time

Fol-lower

Description	Remarks:
Length _____ ft. _____ in.	Weight _____ lb.	

Pene-tration

Elev. of Cutoff	Elev. of Ground
Elev. of Tip	Lg.Cutoff to Tip _____ ft. _____ in.

Ft.	No. of Blows	Speed, Blows Per Min.	Ft.	No. of Blows	Speed, Blows Per Min.	Ft.	No. of Blows	Speed, Blows Per Min.	Ft.	No. of Blows	Speed, Blows Per Min.

Form No. 0230-B

Figure 10.2.40. Sample Pile Driving Record (alternate form).

BITUMINOUS PAVEMENT SAMPLE TRANSMITTAL

Form No. 0261 is to be used when transmitting bituminous pavement samples to the testing laboratory.

The appropriate "SAMPLE OF" block on the form should be indicated with a check "√" or an "X."

This form will be prepared in triplicate. The original is to go with the sample to the laboratory for testing. One (1) copy shall be retained by the Resident Engineer and one (1) copy sent to the Project Manager's office.

Constructioneering Northwest Inc.

BITUMINOUS PAVEMENT SAMPLE TRANSMITTAL

Supplier__________ Report No.__________

Contractor__________ Page _____ of _____

Contract No.__________ Date__________

SAMPLE OF ☐ Pavement ☐ Binder ☐ Sheet Asphalt ☐ Asphaltic Concrete ☐ Other (Explain)__________

SAMPLE NO.__________

DATE PAVEMENT PLACED__________

DATE SAMPLE CUT__________

SAMPLE CUT BY__________

QUANTITY REPRESENTED__________

LOCATION OF SAMPLE

OTHER THAN CITY STREET: LOCATION__________

CITY STREET: LOCATION__________

BETWEEN __________ ST. AND __________ ST. STATION

REMARKS: __________

APPROVED

__________ (Inspector) __________ (Resident Engineer)

Form No. 0261

Figure 10.2.41. Sample Bituminous Pavement Sample Transmittal.

CONCRETE TEST REPORT

Constructioneering Northwest Inc.

CONCRETE TEST REPORT

Order No. _____

Report No. _____

Date Moulded _____ Reported _____
Project _____
At _____
Contractor _____
Reported to _____
Concrete Supplier _____
Quantity Represented _____ Cu. Yds. Specimens made by _____ No. Submitted _____
Location of pour (per information from job-site) _____
_____

Strength requirement _____ psi at 28 days   Mix Design No.: _____

MATERIAL PROPORTIONS USED (Quantities per cubic yard of concrete)
Cement _____ lbs. _____ bags Type _____ Brand _____
Fine Aggregate (SSD) _____ lbs. Source _____ Type: _____
Coarse Aggregate (SSD) _____ lbs. Size _____ Source _____ Type: _____
Coarse Aggregate (SSD) _____ lbs. Size _____ Source _____ Type: _____
Water Total _____ Gals. _____ Gals./bag; Actual Slump _____ In.
Admixture: Amount _____ Kind _____
Admixture: Amount _____ Kind _____ Entr. Air _____ %
Temperatures: Air _____ °F Concrete _____ °F Weather _____
Mixing: Central Mix ☐   Truck Mix ☐   Job Mix ☐
Above data obtained from: CNI Inspection ☐   Project Engineer ☐   Contractor ☐
                          Concrete Supplier ☐   Other ☐ (Designate) _____
Cyls. delivered to laboratory by: CNI ☐   Contractor ☐   Engineer ☐   Common Carrier ☐

COMPRESSIVE STRENGTH

SPECIMEN MARKING	AGE DAYS	LABORATORY NUMBER	DATE RECEIVED	DATE TESTED	TOTAL LOAD–POUNDS	UNIT LOAD – P.S.I.	REPORT NO.

Remarks: _____
_____
_____
_____

_____
Project Manager

Figure 10.2.42. Sample Concrete Test Report.

FIELD DATA ON CONCRETE COMPRESSION
TEST SPECIMENS

Field Data On Concrete Compression Test Specimens

Date Moulded: _____ , 19 ____

Project: _____

At: _____

Contractor: _____

Reported to: _____

Concrete Supplier: _____

Quantity represented _____ Cu. Yds. Specimens made by: _____ No. Submitted: _____

Location of pour _____

Strength requirement _____ psi at 28 days

MATERIAL PROPORTIONS USED (Quantities per cubic yard of concrete) MIX No.:_____

Cement _____ lbs. _____ bags Type _____ Brand _____
Fine Aggregate (SSD) _____ lbs. Source _____ Type: _____
Coarse Aggregate (SSD) _____ lbs. Size _____ Source _____ Type: _____
Coarse Aggregate (SSD) _____ lbs. Size _____ Source _____ Type: _____
Water Total _____ Gals. _____ Gals./bag
Admixture: Amount _____ Kind _____
Admixture: Amount _____ Kind _____
Above information obtained from: CNI batching inspection ☐   Project Engineer ☐   Contractor ☐
                                 Concrete Supplier ☐   Other ☐ (Designate) _____

Entr. Air _____ %   Actual Slump _____ In.   Ticket No.: _____
Temperatures: Air _____ °F Concrete _____ °F Weather _____
Mixing: Central Mix ☐   Truck Mix ☐   Job Mix ☐
Above Site Information obtained from: CNI Inspection ☐   Project Engineer ☐   Contractor ☐
                                      Concrete Supplier ☐ Other ☐ (Designate) _____
Cyls. delivered to laboratory by: CNI ☐   Contractor ☐   Engineer ☐   Common Carrier ☐

SPECIMEN MARKING	TEST AT DAYS	LAB NO.	DATE REC'D.	DATE TESTED	TOTAL LOAD—LBS.	UNIT LOAD—P.S.I.	REPORT NO.
A							
B							
C							
D							
E							
F							

If test results are low, note below if there is any discernible cause.
Remarks: _____

_____

Signed: _____

Figure 10.2.43. Sample Field Data on Concrete Compression Test Specimens.

NON-DESTRUCTIVE TEST REPORT

Constructioneering Northwest Inc.
Engineers

400 Building, 400 108th Avenue N.E. – Bellevue, Washington 98004
Phone (206) 455-9230 or 232-1190

245

Job No _____
Client No _____
Report No _____
Date _____

NON DESTRUCTIVE TEST REPORT

project _____
spec. _____

material _____

Radiographic	Magnetic Particle	EXAMINATION OF :
Radiographic Spot	Penetrant	
Ultrasonic	Other	

identification	remarks			interpretation

No. Films :			HOURS WORKED			INSPECTOR
Film Size			STANDBY			TECH.
			TRAVEL	OTHER :		
			NO MILES	OTHER :		

signature _____

Figure 10.2.44. Sample Non-Destructive Test Report.

**INSPECTION ACTIVITY SHEET
AND
CONCRETE POUR CARD**

INSPECTION ACTIVITY SHEET & CONCRETE POUR DATA

Contract Title_____ Job No. _____

POUR DATA

Concrete Pour No._____   Scheduled Pour Date_____

Location_____ Elevation: From_____ To_____

Actual Pour Date_____ Time: Start_____ Finish_____ Mix No._____

Type Mix_____ Estimated Qty. - C.Y._____ Actual Qty._____

Waste - C.Y.:_____   DELIVERY TICKET NUMBERS

Foreman:_____

Inspector:_____

Weather:_____ Temp.:_____

REMARKS:_____

CLEARANCE DATA

ITEM	REQUESTED INSPECTION #1	#2	#3	REMARKS
Excavate & Subgrade				
Forms - Line & Grade				
Forms - Braced, Ready to Pour				
Re-steel - Quantity				
Re-steel - Secure in Place				
Pipe - Quantity - Secure				
Electrical Conduit - Quantity - Secure				
Ground Wire				
Anchor Bolts & Rods				
Miscellaneous Metal				
Green Cut - Subject to Proper Care of Joint				
Final Cleanup				
O.K. to Pour				

(Use Reverse Side for Additional Notes, if Required)

Form No. 0330

Figure 10.2.45. Sample Inspection Activity Sheet and Concrete Pour Data.

REPORT OF ULTRASONIC EXAMINATION
OF WELDED GIRDERS

400 Building, 400 108th Avenue N.E. - Bellevue, Washington 98004
Phone (206) 455-9230 or 232-1190

ENGINEERS
CONSTRUCTION CONSULTANTS

DATE_____

REPORT OF
ULTRASONIC EXAMINATION OF WELDED GIRDERS

PROJECT: _____

SPECIFICATION: _____

FABRICATING SHOP: _____ SHOP CONTR. NO. _____ SHOP DWG. NO. _____

REPORTED TO: _____

							DATA DESCRIBED IN AWS D2.0-69 APPENDIX C								
DATE	WELD LOCATION	ACCEPTABLE	REJECTABLE	TRANSDUCER ANGLE	MODE	DECIBELS				ANGULAR (DISTANCE SOUND-PATH)	LENGTH	DEFECT			
						DEFECT LEVEL	REFERENCE LEVEL	ATTENUAT FACTOR	DEFECT RATING			DEPTH "A" FROM SURFACE	DISTANCE FROM WELD "X" CENTER	DISTANCE FROM WELD-END "Y" (STAMPED)	
						A	B	C	D						

WELD LOCATION AND IDENTIFICATION SKETCH

ERECTION
MARK

_____   Project Manager

WITNESSED   _____   _____

Figure 10.2.46. Report of Ultrasonic Examination of Welded Girders.

SIEVE ANALYSIS REPORT

Constructioneering Northwest Inc.

**ENGINEERS
CONSTRUCTION CONSULTANTS**

SIEVE ANALYSIS

LABORATORY NO. _____

PROJECT _____   CLIENT'S NO. _____

LOCATION _____   DATE _____

_____ AGGREGATE _____ %   _____ AGGREGATE _____ %

SAMPLE NO.					SAMPLE NO.				
U.S. STD. SIEVE	WEIGHT RETAINED	% RETAINED	% PASSING	SPECIFI-CATIONS	U.S. STD. SIEVE	WEIGHT RETAINED	% RETAINED	% PASSING	SPECIFI-CATIONS
TOTAL					TOTAL				
WEIGHT ORIGINAL SAMPLE					WEIGHT ORIGINAL SAMPLE				

_____ AGGREGATE _____ %

SAMPLE NO.				
U.S. STD. SIEVE	WEIGHT RETAINED	% RETAINED	% PASSING	SPECIFI-CATIONS
TOTAL				
WEIGHT ORIGINAL SAMPLE				

REMARKS:

REMARKS:

TESTED BY _____

Figure 10.2.47. Sample Sieve Analysis Report.

REPORT OF IN-PLACE SOIL DENSITY TESTS

Order No. _____
Report No. _____ _____

REPORT OF IN-PLACE SOIL DENSITY TESTS

Client_____

Project_____

Soil Description _____

Max. Dry Density_____ lbs./cu. ft. Optimum Moisture_____ % Method of Test_____

DATE OF TEST	TEST NO.	TEST LOCATION	ELEV. FT.	LIFT NO.	FIELD MOISTURE %	IN-PLACE DENSITY (LBS./CU. FT.)		% COMPACTION
						WET	DRY	

Remarks_____
_____
_____
_____
_____
_____

Project Manager

_____

Figure 10.2.48. Sample Report of In-Place Soil Density Tests.

PHYSICAL TEST REPORT

ENGINEERS
CONSTRUCTION CONSULTANTS

WELDING - PHYSICAL TEST REPORT   DATE _____

_____ QUALIFICATION TEST
PROCEDURE OR WELDER

For_____

Specimens Welded In Accordance With Spec. _____
PROCESS SPECIFICATION

Material _____ Thickness _____ Spec. No. _____
(PIPE OR PLATE)

Was Backing Strip Used? _____   Single or Double Weld _____

Welders Name _____ Stamp No. _____

Was Welding Witnessed by Representative of CNI?   Yes ☐   No ☐

If "Yes", Witnessed By _____ Date_____ Where Welded _____

TEST RESULTS

SPEC. NO.	POSITION	WIDTH	THICKNESS	AREA	TOTAL LOAD (POUNDS)	ULTIMATE UNIT STRESS PSI	FRACTURE	PASSED OR FAILED

SPEC. NO.	POSITION	TYPE OF BEND	RESULTS	SPEC. NO.	POSITION	TYPE OF BEND	RESULTS

Project Manager

_____

Figure 10.2.49. Sample Physical Test Report form.

REPORT OF MOISTURE-DENSITY
RELATIONSHIP OF SOIL

ENGINEERS
CONSTRUCTION CONSULTANTS

Date _____

MOISTURE - DENSITY RELATIONSHIP OF SOIL

PROJECT _____

Location_____

Sample_____

Method of Test_____

MOISTURE - DENSITY RELATIONSHIP CURVE

DRY DENSITY LBS./CU.FT.

MOISTURE CONTENT %

Optimum Moisture_____% Max. Dry Density_____lbs./cu.ft.

Project Manager

_____ ____

Figure 10.2.50. Sample Report of Moisture-Density Relationship of Soil.

VIII. SAFETY

A. GENERAL POLICY

The general policy of the AECM team related to Safety is to be that every employee is responsible for both his own personal safety and that of others on the project. In the event unsafe conditions are observed, these observations shall be immediately brought to the attention of the responsible party.

The Resident Engineer is to insure that acceptable safety standards are observed on the work. In doing this, he will manifest an aggressive, sincere interest in the Safety Program. To this end, the Resident Engineer and his staff shall:

1. Familiarize themselves with the safety requirements, safety standards, and codes applicable to the construction operation.
2. Assist AECM team subcontractors or prime contractors and their subcontractors in analyzing and resolving job safety problems.
3. Make safety inspections, if necessary or appropriate.
4. Bring any apparent safety violations to the attention of the responsible people involved, requesting prompt correction.

In the event a subcontractor or a contractor is slow, or refuses to comply with a request for correction of conditions contrary to safety requirements, the Project Manager and/or Resident Engineer shall be notified. No action under the Stop-Order provision or similar provisions of the contract shall be taken except by the Project Manager or Resident Engineer unless the problem is of extreme emergency nature.

All major accidents shall be reported to the Resident Engineer and to the Project Manager.

In general, the Contract Documents will provide that each contractor and/or subcontractor is responsible for setting up and maintaining his own safety program. At the start of a contract, the Resident Engineer is to request that he be furnished the details of this safety program by the contractor/subcontractor.

Items that should be covered in this safety program are:

1. Employee safety indoctrination and instruction.
2. Maintenance of interest in Safety.
3. Participation by subcontractors.
4. Tool Box Meetings.
5. First Aid and Medical facilities.
6. Action in emergencies.
7. Fire prevention and protection.
8. Inspection of unsafe working conditions.

9. Housekeeping practices.
10. Inspection of tools and equipment.
11. Equipment operation.
12. Public protection.
13. Sanitation.
14. Accident reporting and claim handling.

If a contractor or subcontractor does not follow his Safety Program, the Resident Engineer must discuss the situation with the responsible contractor's or subcontractor's representative.

In addition to the above, the Resident Engineer shall insure that Safety and Accident Reports, if required under the contract(s), are submitted, and that all safety requirements are satisfied.

Sample forms that may be useful in establishing and maintaining a Safety Program and its records follow. It is suggested that the Resident Engineer adopt these and/or other forms, as appropriate, and have adequate files of them maintained in his offices.

B. CONTRACTOR-COMPLETED ACCIDENT REPORTS

1. Contract specifications will usually provide that the contractor is to complete and furnish to the Resident Engineer the forms referred to in Section 2, following (or equivalent forms).

2. At the Preconstruction Conference, or by letter prior to the start of work on any contract, the Resident Engineer shall request that the contractor or subcontractor provide him with two (2) copies of each of the following forms. At the same time, the Resident Engineer shall furnish the contractor or subcontractor with the following instructions pertaining to each form.

   a. Contractor's Accident Report to His Insurance Carrier
   This form is to be submitted for each accident reported by the contractor.

   b. Supervisor's Report of Accident
   A Supervisor's Report of Accident, Form No. 0140, is to be submitted by the contractor for each accident resulting in lost time to personnel, including persons not directly connected with the project, and for accidental damage to property or equipment. Submittal shall be made as soon as possible, but in no case later than forty-eight (48) hours after the accident. Pertinent facts which are not available within the above-mentioned time shall be submitted as soon as available in a supplemental report. (See Figure 10.2.51 for a sample of Form No. 0140.)

```
╭─────────────────────────────────────────────────────────────────────╮
│  ╭Constructioneering╮                                                 │
│  │    Northwest     │          SUPERVISOR'S REPORT OF ACCIDENT        │
│  │      Inc.        │                                                 │
```

1. Contract No.	2. Accident Date	3. Time of Accident	4. Project Section

5. Location of Accident	6. Reporting Organization	7. Contractor Involved

8. Injury ☐ Lost Time ☐ Critical ☐ Fatal	9. Damage* ☐ Fire ☐ Property ☐ Equipment

10. Injured Person & Address	11. Occupation of Injured Male ☐ Employer & Address Female ☐ Age____

12. Nature of Injury	13. Date Stopped Work	14. Date Returned

15. First Aid By	16. Ambulance

17. Hospital	18. Attending Physician

19. Witnesses or persons responding, including addresses

20. Fire Department	21. Police Department

22. Equipment and/or materials involved

23. Primary cause of accident

24. Secondary cause of accident

25. Contributing factors

26. Supervisor's Corrective Action

_____
(Supervisor's Signature)

27. Project Superintendent's Corrective Action

_____
(Project Superintendent's Signature)

28. Date of This Report
_____

* Attach a list of damaged property and/or equipment, excluding motor vehicles.
  Indicate names and addresses of equipment or property owners.

Form No. 0140

Figure 10.2.51.A. Sample form for Supervisor's Report of Accident
(front page).

SUPERVISOR'S REPORT OF ACCIDENT

(CONTINUATION)

29. Safety Regulations Involved: Chapter   Paragraph

30. Photographs attached

31. Sketch showing location of nearby structures, materials, equipment, etc.,
    with approximate scale of distances.

32. Narrative description of events immediately prior to, during and immed-
    iately after the accident.

Figure 10.2.51.B. Sample form for Supervisor's Report of Accident
(reverse side).

c. Accident Experience Breakdown

An Accident Experience Breakdown report, Form No. 0141, is to be submitted by the contractor and his subcontractors to reflect a separate monthly tabulation of accidents involving his organization and each of his subcontractors' organizations. Each accident resulting in lost time to personnel, including persons not directly connected with the project, and accidental damage to property or equipment, shall be listed separately and numbered consecutively by each contractor or subcontractor. (See Figure 10.2.52 for sample.)

d. Accident Experience Summary

An Accident Experience Summary report, Form No. 0142, is to be submitted periodically by the contractor and his subcontractors. Since this form is a cumulative total of all accidents reported on Form No. 0141, it thereby reflects the total project accident experience of each contractor and subcontractor. (See Figure 10.2.53 for a sample of this summary form.)

e. Construction Safety Survey

(1) A construction Safety Survey, Form No. 0150 (see Figure 10.2.54 for sample), may be required periodically from contractors and subcontractors as a permanent record of actions taken to reduce potential hazards and to insure that the contractors and subcontractors are implementing their Safety Programs. At the discretion of the Resident Engineer, requirements for this form may be deleted or modified, provided written records of an adequate contractor Safety Program are obtained.

(2) Form No. 0150 is primarily intended for preparation by the contractor's representative who is appointed to perform safety inspection services in accordance with contract specifications. The Resident Engineer shall use it to notify the contractor of certain hazards brought to his attention but not previously corrected. The Construction Safety Engineer may use it to report his recommendations made during his periodic safety survey of the project.

(3) The contractor's representative may prepare the form as necessary, but in no event later than Friday of each week; however, potential hazards involving imminent danger shall be corrected immediately and duly reported on this form. Reports prepared by the Resident Engineer or the Construction Safety Engineer shall be followed up for immediate corrective action. The contractor's weekly report

Constructioneering Northwest Inc.

ACCIDENT EXPERIENCE BREAKDOWN

Page _____ of _____

CONTRACT NO. _____

PROJECT

	MANHOURS WORKED	LOST TIME INJURIES	DAYS LOST	FREQUENCY RATE	SEVERITY RATE	ESTIMATED DAMAGE COST
Previous Total						
This Month						
Total To Date						

ACCIDENT REPORT NO.	DATE OF ACCIDENT	NAME OF INJURED	DAYS LOST THIS MONTH	F = Fire Damage P = Property Damage E = Equipment Damage	ESTIMATED DAMAGE COST

Reporting Organization

Report Prepared By _____
    (Name)   (Title)

Signature of Project Supt. _____

Month – Year

Form No 0141

Figure 10.2.52. Sample form for Accident Experience Breakdown.

REPORTING ORGANIZATIONS	MANHOURS WORKED TO DATE	LOST TIME INJURIES TO DATE	DAYS LOST TO DATE	FREQUENCY RATE TO DATE	SEVERITY RATE TO DATE	ESTIMATED DAMAGE COST TO DATE

ACCIDENT EXPERIENCE SUMMARY

Page _____ of _____

CONTRACT NO.

Constructioneering Northwest Inc.

PROJECT

TOTAL PROJECT EXPERIENCE

Reporting Organization

Report Prepared By

(Name)   (Title)

Signature of Project Supt.

Month – Year

Form No. 0142

Figure 10.2.53. Sample form for Accident Experience Summary.

Constructioneering Northwest Inc.

CONSTRUCTION SAFETY SURVEY

Contract Number____________ Report Number__________

Contractor____________

ITEM NO.	RECOMMENDATION	SAFETY REGULATION REFERENCE	ACTION TAKEN AND/OR COMPLETED

SURVEY MADE BY:

SIGNATURE:____________

TITLE:____________

DATE:____________

SIGNATURE OF CONTRACTOR'S PROJECT MANAGER ____________ DATE ______

SIGNATURE OF RESIDENT ENGINEER ____________ DATE ______

Form No. 0150

Figure 10.2.54. Sample form for Construction Safety Survey.

shall reflect all hazards noted formally or informally during the week, including the date of corrective action.

3. As an alternative to the reports outlined in Section 2-a through 2-e, above, the Resident Engineer and contractor may, with permission from the Project Manager, set up a new procedure wherein applicable and equal reports—such as those of the contractor's insurance carrier or OSHA—are furnished, provided the same or equal information is obtained.

C. AECM TEAM-COMPLETED SAFETY REPORTS

1. General

   The Resident Engineer shall be responsible for the AECM team's Safety Program and all safety reports.

   As a minimum, the two reports discussed in Section 2, below, shall be completed and filed in the offices of the Resident Engineer. This procedure will be augmented by the Resident Engineer, as appropriate.

2. Reports Required—Minimum
   a. REPORT OF PERSONAL INJURY
      (1) A Report of Personal Injury, Form No. 0131, will be prepared by the field forces to report each and every injury, no matter how slight. (See Figure 10.2.55 for a sample of this reporting form.)
      (2) Form No. 0131 will be completed at the time first aid is given, or as soon after the personal injury as possible, circumstances considered. In no case shall the Report of Personal Injury form be submitted later than the day immediately following the date of the injury.
      (3) This form will be completed in triplicate by the Resident Engineer or the Construction Safety Engineer. Two (2) copies shall be submitted to the Project Manager immediately upon completion.
   b. REPORT OF ACCIDENT OR DAMAGE TO EQUIPMENT OR PROPERTY
      (1) This form, No. 0132, will be prepared to cover each and every accident to equipment or property, no matter how slight. (See Figure 10.2.56 for a sample of the Report of Accident or Damage to Equipment or Property form.)
      (2) Form No. 0132 will be prepared from information resulting from investigation or direct reports of the person or persons involved in or responsible for the accident.

```
Constructioneering
Northwest
Inc.
                                    REPORT OF PERSONAL INJURY

CONTRACTOR_____      DATE OF REPORT_____
                                       CONTRACT NO._____
SUBCONTRACTOR_____      SECTION DESIGNER_____
                                       REPORT NUMBER_____

NAME OF INJURED_____      SEX: M ☐   F ☐   AGE:____
ADDRESS_____      MARRIED ☐  SINGLE ☐
OCCUPATION_____      NO.DEPENDENTS___ AGES:____

LOCATION WHERE INJURY OCCURRED_____

TIME OF INJURY: _____AM _____PM   ACCIDENT DATE_____
WEATHER _____ TEMP:_____°F

TYPE OF INJURY:_____

HOW SUSTAINED:_____
_____
_____

WAS INJURED PERSON TAKEN TO: DOCTOR? ☐  HOSPITAL? ☐ OTHER? ☐*
*IF OTHER, EXPLAIN HERE:    _____

NAME & ADDRESS OF DOCTOR:   _____

NAME & ADDRESS OF HOSPITAL: _____

WILL INJURED PERSON LOSE TIME? YES ☐ NO ☐ EST.LOST TIME_____
WAS SAFETY RULE ALLEGEDLY VIOLATED? YES ☐   NO ☐
IF YES, WHAT RULE WAS ALLEGEDLY VIOLATED?_____

ACTION TAKEN TO PREVENT REPETION_____
_____

WITNESSES TO ACCIDENT       WAS STATEMENT TAKEN FROM INJURED?
_____       YES ☐   NO ☐
_____      WAS STATEMENT TAKEN FROM WITNESS(ES)?
_____       YES ☐   NO ☐
_____
_____      Note: Use back of this sheet for
_____      Statement(s) taken, additional
                            explanation, or sketches.

        SIGNED_____ TITLE_____
Form No.0161
```

Figure 10.2.55.A. Sample Report of Personal Injury form (front page).

REPORT OF PERSONAL INJURY

(CONTINUATION)

MAKE DETAILED STATEMENT AS TO CAUSE AND RESULTS OF ACCIDENT:

_____

_____

_____

_____

_____

_____

_____

USE THIS SPACE FOR ADDITIONAL EXPLANATION OR SKETCHES:

Figure 10.2.55.B. Sample Report of Personal Injury form (reverse side).

```
┌──────────────────────────────────────────────────────────────────┐
│ Constructioneering                                                 │
│   Northwest              REPORT OF ACCIDENT OR DAMAGE              │
│     Inc.                    TO EQUIPMENT OR PROPERTY               │
│                                                                    │
│                          Date of Report_____             │
│ CONTRACTOR_____  Contract No._____      │
│                                  Section Designer_____      │
│ SUBCONTRACTOR                    Report No._____      │
│                                                                    │
│ LOCATION OF ACCIDENT_____        │
│                                                                    │
│ EQUIPMENT INVOLVED (DESCRIPTION, SERIAL NOS., OWNER)_____      │
│                                                                    │
│ DAMAGE RESULTING FROM ACCIDENT_____        │
│                                                                    │
│ WERE THERE PERSONAL INJURIES?  YES ☐* NO ☐  *If "Yes", prepare     │
│                                               Form No. 0161.       │
│ ESTIMATED VALUE OF DAMAGES: $_____                        │
│                                                                    │
│ WITNESSES TO ACCIDENT      WAS (WERE) STATEMENT(S) OBTAINED         │
│ _____   FROM WITNESS(ES)?                       │
│ _____       YES ☐    NO ☐                       │
│ _____   IS (ARE) STATEMENT(S) ATTACHED?         │
│ _____       YES ☐    NO ☐                       │
│ _____                                           │
│ _____                                           │
│ _____                                           │
│                                                                    │
│ REMARKS:_____         │
│ _____         │
│ _____         │
│ _____         │
│                                                                    │
│ TIME OF ACCIDENT: _____AM _____PM   DATE_____         │
│ WEATHER CONDITIONS_____   TEMPERATURE_____°F         │
│ ROADWAY OR SURFACE CONDITIONS:                                     │
│    WET ☐      DRY ☐      ICY ☐       OTHER ☐ *                     │
│ *IF "OTHER", EXPLAIN:_____        │
│ _____         │
│ _____         │
│                                                                    │
│     If more space is required, use back of this sheet for addi-    │
│     tional information, statements, or sketches.                   │
│                                                                    │
│        SIGNED_____  TITLE_____         │
│ Form No. 0162                                                      │
└──────────────────────────────────────────────────────────────────┘
```

Figure 10.2.56. Sample Report of Accident or Damage to Equipment or Property.

(3) Form No. 0132 will be completed within forty-eight (48) hours of the time of the accident.
(4) The form will be completed in triplicate by the Resident Engineer. Two (2) copies shall be submitted to the Project Manager immediately upon completion.

Chapter 11

Putting It Together

In this last chapter, two diverse subjects serving as cohesive elements in AECM work—Cost Engineering and Communications—will be discussed.

COST ENGINEERING

The work of Cost Engineering includes preparation of order-of-magnitude or conceptual estimates, preliminary budgets and estimates, control estimates, forecasts and trend reports, and comparative estimates. Cost studies are included, as well as cost control programs and coordination of these, cash flow studies, engineer's estimates, "Hard money" estimates, analysis of bids, change order evaluations, and contractor's claim studies. Information and costs for negotiations and participation in, or conducting of, negotiations, if necessary can also be part of cost engineering, as can value engineering studies.

Since estimating and cost controlling are principal among these functions, and because of space limitations, we will limit our discussions to these portions of cost engineering.

A principal objective in AECM work is the creation of constructed work which will meet functional and aesthetic requirements within budget. Because of this objective, estimating and cost controlling work is necessary at every phase of a project—from concept through completion of construction.

The development of an estimating and cost control program early and implementation of this program throughout a project's life is mandatory in all AECM work. As a minimum, this program, prior to bidding or construction, should include: preparation of a budget, cost approximation or preconceptual estimate, and cost control system; preparation of an estimate at the end of planning—conceptual and schematic phase and improvement of previously established cost controls; preparation of at least two estimates during design development; and preparation of an estimate prior to completion of Contract Documents phase.

As early as feasible in the project, determinations should be made as to: the Owner's basic philosophy and objectives as they relate to the project; the amount of money available or the overall budget; the space and area requirements; and the possible space and area relationships.

With this information as background, a cost control breakdown should be developed to indicate amounts available, such as:

Cost Control Area	Dollars
1. Demolition Site & Civil Work	A
2. Structural Work	B
3. Architectural Work	C
4. Electrical Work	D
5. Mechanical Work	E
6. Vertical Transportation Work	F
7. Other Expenses	G
8. Contingencies	H
TOTAL	X

With this information, the designers in all disciplines can develop a realistic feel for the project and the degree of refinement or cost control that may be necessary. The importance of making this type of distribution early is evident.

From this point on throughout the project, Cost Engineering work is a process of reporting, refining, and fine-tuning of cost controls, budgets, and estimates.

AECM cost estimates are usually prepared by either the unit cost pricing method or the crew-materials-equipment pricing method or, preferably, through the use of both of these methods.

The unit cost pricing method is favored in building and specialty estimating and is valuable in AECM work for preliminary estimates. In this method, items of work are organized, listed, and quantified, then item or unit prices are assigned for each work item.

The crew-materials-equipment method of estimating is favored in and is readily applicable to heavy construction estimates. In this method, the construction crew, crew and equipment, or crew and equipment plus materials to be employed are estimated and costed for the operation established and the period of time required to complete the item or to produce a specific amount of work. This cost is then used to project the estimated total cost.

The end products resulting from either of these methods are accurate cost estimates based on reasonable judgment and reliable information.

The author has found in AECM work that, by using tabular forms containing complete listings of all items in the project, costly and/or embarrassing errors can be reduced. Additionally, these forms can be made to indicate vividly to all concerned parties where the construction dollars are going.

As an example of this use of tabular forms, consider an estimate prepared at the end of the conceptual or schematic phase of design that can, and possibly should, include listed parameters such as type and class of structure(s); total floor and building area; type and quantity of structural materials (possibly based on projection of partial design); type, size, and quantity of mechanical systems and work; type, size, and quantity of electrical systems and work; number and types of vertical transportation systems; type, size, and quantity of architectural finishes; and type, size, and quantity of site, civil, and foundation work.

At the end of the conceptual or schematic design, it may not be possible to estimate, from quantity information, budgets based on size (such as so much per square foot of floor area), but if at all possible estimates should be based on detailed quantitative information.

Every estimate prepared after the conceptual or schematic estimate should be increasingly more detailed.

For convenience of reference, a sample concept estimate with estimating information "parameters" is shown in Figures 11.1 and 11.2.

Although this early estimate is broad in scope, it does take a first step in the establishment of cost controls and design criteria. Each subsequent estimate should increase in detail and, with this increased detail, should provide for a reduction in contingencies.

It must be recognized that, in the early stages of design, two items are extremely difficult to estimate—contingency and escalation. The amount allowed for contingencies should usually be a function of the experience and confidence estimate. If it is determined that the estimate is based on similar past projects, or what could be called "safe grounds," contingency should be somewhere near the following percentages:

Preschematic or concepts	25% ±
Concepts	15% ±
Preliminary Design Development	12%
Prefinal and Final	5 *to* 10%

In the prefinal stages, this contingency amount can often serve as an allowance against extras and changes as well as a cushion against high bids. The refinement of this should be a function of project and client requirements.

Escalation costs are real and can be estimated and are most important in projects taking a period of years to construct. References and projections

Construction•meeting Northwest Inc.

ESTIMATE SUMMARY

LIBRARY FACILITIES – UNIVERSITY
PROJECT

CONCEPT REVIEW MEETING
PREPARED FOR

PREPARED BY J. Jones
CHECKED BY S. Smith

15 July 1978
DATE
SHEET 1 OF 2

NO.	DESCRIPTION	QUANTITY	UNIT	UNIT COSTS	AMOUNT	TOTALS	REMARKS
	ESTIMATING PARAMETERS						NOTE – Nearing completion of conceptual design – space limits still not set.
A	Area Per Floor	200 x 150			30,000 SF		
B	4 Floors	4 x 30,000			120,000 SF		
C	Usable Area				101,000 SF		
D	Materials Description – Highest Quality						
	ESTIMATE DETAILS –						NOTE – See Est. Detail Shts.
	Demolition	1	Bldg		10,000	10,000	
	Clearing & Grubbing	3	Acre		2,500	7,500	
	Excavation	30,000	CY	3.00	90,000	90,000	
	Backfill	6,000	CY	5.00	30,000	30,000	
	Sub Structure Walls	8,400	SF	5.00	42,000	42,000	
	Footings & Grade Bms.	3,100	CY	180.00	558,000	558,000	
	Caissons	500	CY	100.00	50,000	50,000	
	Shoring Exc.	6,000	SF	20.00	120,000	120,000	
	Structural Steel	250,000	Lb	2.50	625,000	625,000	
	Conc. Slab on Grade	30,000	SF	1.50	45,000	45,000	
	Conc. Sup. Slab	120,000	SF	4.40	528,000	528,000	
	Misc. Metal	110,000	Lb	2.00	220,000	220,000	
	Fireproofing		LS			40,000	
	Floor Finish					255,000	
	Conc.	30,000	SF	0.50	15,000		
	Vat	60,000	SF	1.00	60,000		
	Rug	60,000	SF	3.00	180,000		
	Suspended Ceiling	120,000	SF	1.10	132,000	132,000	
	Partitions	180,000	SF	3.00	540,000	540,000	
	Painting	500,000	SF	0.25	125,000	125,000	
	SUB-TOTAL, PAGE 1					3,417,500	(cont. on page 2)

Figure 11.1. Sample Concept Estimate Summary (page 1).

PREPARED BY J. Jones
CHECKED BY S. Smith

ESTIMATE SUMMARY

LIBRARY FACILITIES – UNIVERSITY
PROJECT

CONCEPT REVIEW MEETING
PREPARED FOR

15 July 1978
DATE

SHEET 2 OF 2

NO.	DESCRIPTION	QUANTITY	UNIT	UNIT COSTS	AMOUNT	TOTALS	REMARKS
	Superstructure and Walls						*General Conditions incl. in unit prices at this stage.
	Perimeter 700 ft.						
	Area 700 x 60	42,000	SF	---	Below		
	Proportions based						
	Elev. view						
	Brick & Block work	5,000	SF	6.00	30,000		
	Curtain Wall	30,000	SF	10.00	300,000		
	Insulated Glass	7,000	SF	15.00	105,000		
	Doors & Hardware						
	Entry, Main	3	Ea	3,000.	9,000		
	Entry, Service	2	Ea	2,000.	4,000		
	Std. & Frames	65	Ea	300.00	19,500		
	Stairs	12	Flr	1,500.	18,000		
	Entry	2	Flr	3,000.	6,000		
	Roofing Insulation & Flashing	30,000	SF	1.00	30,000		
	Mechanical	120,000		8.00	960,000		
	Electrical	120,000		5.00	600,000		
	Elevators	Allowance			100,000		
	Site	Allowance					
	SUB-TOTAL, PAGE 2				2,181,500		
	SUB-TOTAL, PAGE 1				3,417,500		
	SUB-TOTAL, BOTH PAGES				5,599,000		
	Design Allowance & Contingencies				1,001,000		
	TOTAL BUDGET				6,600,000		

Figure 11.2. Sample Concept Estimate Summary (page 2).

developed from past bids and from published cost indices are valuable in estimating escalation costs.

Despite all these facts—the competence of estimators and cost engineers, information derived from past records, etc.—in AECM work as well as construction work, estimating is an imprecise art. This explains in part the reason for the diversity in bids—which can vary as much as 10 to 15 percent—from different contractors on the same project.

It is, therefore, all-important that the AECM team keep the client informed. Facts such as the following should be conveyed: costs rise from 3 percent to 18 percent per year due to labor rate increases, thus delay of a "go-ahead" on a project will increase costs; "crash" construction programs add to costs and, therefore, it is advisable to allow contractors sufficient time for construction; weather and seasons affect costs; allowances must be made for design development; and programs must be adhered to or costs can go up. The quality of any estimate is a function of the time spent estimating and also of the sufficiency of information. One hour is not sufficient for a $15 million project estimate. A detailed estimate cannot be made adequately from a preliminary sketch or rendering.

Further, higher bids can be caused by: lack of communication between the client and the AECM team; scope changes—if the project grows during design or as a result of client changes, it will likely cost more; shortening the duration of the construction; limiting competition, through specifying single sources or products; overspecifying—e.g., requiring stainless steel or special grade steel when standard steel would do; late changes—a 100-page addendum or six 10- to 20-page addenda will certainly make contractors leery of the design and prone to increase their bid price.

Lastly, in this brief discussion of estimating, it seems appropriate to mention Contractor General Condition and Plant Costs—an item that should be detailed in all more final estimates.

A sample format, which can be used advantageously to identify all of these elusive items of costs for AECM estimates, is shown in Figure 11.3.

The point to remember is that, through subdividing and evaluating costs, and because of the efforts of compensating errors, the more detailed and divided the final estimates are the more likely they are to be right.

A discussion of Communications follows the Contractor General Condition and Plant Costs forms (Figure 11.3).

COMMUNICATIONS

The needed flow of cost, design, and other information within and outside the AECM team gives rise to the following comments.

Constructioneering
Northwest
Inc.

**ENGINEERS
CONSTRUCTION CONSULTANTS**

Est. No._____
Date:_____
Sht.No._____ of _____
By:_____ Ckd:_____

Project:_____ Location:_____
Owner:_____ Designer:_____
Bid Time:_____ Bid Place:_____ Contract Time:_____
Liquidated Damages:_____

	LABOR	MATERIAL	EQUIPMENT	TOTAL
LABOR & EXPENSE				
Indirect Personnel				
Indirect Expense				
Temp. Facilities				
EQUIPMENT				
Ownership & Supplies				
Small Tools				
Move In & Out				
Rentals				
SUB-TOTAL				
G.C. LABOR T & I				
G.C. LABOR FRINGES				
TOTAL GEN. COND.				
SPECIAL ITEMS				
Labor Escalation				
Mat'l. Escalation				
O.T. Allowances				
Shift Premium				
GRAND TOTAL				

Figure 11.3A Summary—General Conditions & Plant.

```
                                    Est.No. _____
                                    Date: _____
                                    Sht.No. ____ of ____
                                    By: _____ Ckd: _____
```

INDIRECT PERSONNEL (1)	QTY.	UNIT	UNIT L	LABOR	UNIT M	MATERIAL	UNIT T	TOTAL
SUPERVISORY STAFF								
Project Manager		MOS						
General Supt.		MOS						
Asst. Supt.		MOS						
Earthwork Supt.		MOS						
Concrete Supt.		MOS						
Architectural Supt.		MOS						
Master Mechanic		MOS						
Master Electrician		MOS						
E.E.O. Officer		MOS						
ENGINEERING STAFF								
Project Engineer		MOS						
Office Engineer		MOS						
Cost Engineer		MOS						
Field Engineer		MOS						
Field Party		MOS						
Draftsman - Detailer		MOS						
Safety Engineer		MOS						
Quality Controller		MOS						
Q.C. Assistants		MOS						
OFFICE STAFF								
Office Manager		MOS						
Purchasing Agent		MOS						
Accountant		MOS						
Payroll Clerk		MOS						
Stenographer		MOS						
Typist		MOS						
Clerk		MOS						
Receptionist		MOS						
SUB-TOTAL								

Figure 11.3B Indirect Personnel Costs (page 1).

Est.No._____
Date:_____
Sht.No._____ of _____
By:_____ Ckd:_____

INDIRECT PERSONNEL (2)	QTY.	UNIT	UNIT L	LABOR	UNIT M	MATERIAL	UNIT T	TOTAL
FIELD STAFF								
Warehouse,Yard,Parts		MOS						
Watchman, Guard		MOS						
Waterboy		MOS						
Teamster		MOS						
Nurse, First Aid		MOS						
LOGISTIC EXPENSE								
Moving								
Travel								
Subsistence								
TOTAL INDIRECT PERSONNEL								

Figure 11.3C Indirect Personnel Costs (page 2).

Est.No._____
Date:_____
Sht.No._____ of _____
By:_____ Ckd. _____

INDIRECT EXPENSE (1)	QTY.	UNIT	UNIT M&S	MAT/SUB	UNIT E	EQUIPMENT	UNIT T	TOTAL
OFFICE								
Heat – Light – Water								
Telephone								
Postage – Stationery								
Furniture								
Machines								
Engineering Supplies								
Reproduction								
Safety Supplies								
Public Relations								
Contributions & Dues								
Entertainment								
CONSULTANTS & SERVICES								
Dewater, Shore, Blast								
Critical Path								
Curvey & Layout								
Testing, Concrete								
Testing, Steel								
Testing,								
Legal								
Audit								
Payroll								
Medical								
Photographs								
Parking								
Executive Travel								
Badges								
Housing & Feeding								
PERMITS & FEES								
State License								
City, County License								
Building Permits								
Special Permits								
Rights of Way								
Royalties								
Explosives								
SUB-TOTAL								

Figure 11.3D Indirect Expenses (page 1).

(Fig. 11.3E)   Est.No._____
                Date:_____
                Sht.No._____of_____
                By:_____Ckd._____

INDIRECT EXPENSE (2)	QTY.	UNIT	UNIT M	MATERIAL			UNIT T	TOTAL
SUB-TOTAL (BR'T. FWD.)								
TAXES								
Payroll								
Sales and/or Use								
Business & Occupation								
Personal Property								
Real Estate & School								
Duties								
Special City								
Special County								
Special State								
INSURANCE & BONDS								
Payroll								
Builder's Risk								
Equipment Floater								
Railroad								
Fidelity								
Fire & Ext. Coverage								
Performance Bond								
Subcontract Bond								
TOTAL INDIRECT EXPENSE								

Figure 11.3E Indirect Expenses (page 2).

Est.No. _____
Date: _____
Sht.No. _____ of _____
By: _____ Ckd. _____

TEMPORARY FACILITIES (1)	QTY.	UNIT	UNIT L	LABOR	UNIT M or S	MAT/ SUB	UNIT T	TOTAL
TEMPORARY BUILDINGS								
Contractor's Office								
Owner's Office								
Change House								
Yard & Sheds								
Warehouse Storage								
Toilets - Job Built								
Toilets - Portable								
DRINKING WATER								
Supply								
Drinking Fountains								
Ice								
Igloos, Cups, etc.								
WATER								
Service - Supply								
Distribution System								
Utility Cost								
ELECTRICAL								
Service								
Distribution System								
Utility Cost								
COMPRESSED AIR								
Install Equipment								
Distribution System								
Operating Engineer								
SUB-TOTAL								

Figure 11.3F Temporary Facilities Costs (page 1).

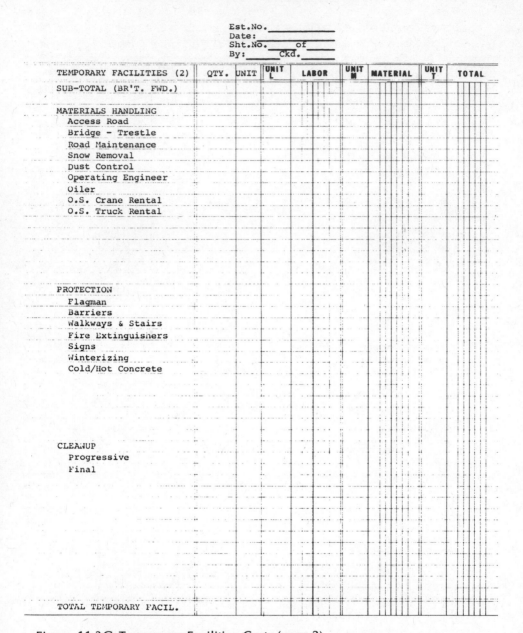

Figure 11.3G Temporary Facilities Costs (page 2).

Effective communication can be either a blessing or a problem in AECM work. To insure that communications do not become a problem, those involved in AECM work must learn to pay more attention to what they say and how they say it—and to what they do.

Messages, procedures, directives, memos, and letters must be prepared in such a manner as to be clear, concise, and easily understood; leave no question in the mind of the recipient as to what is meant.

In AECM work, informal communication is desirable; nevertheless, in this type of work it is impossible and impractical to rely on informal or word-of-mouth communication exclusively. Something must be maintained as a record.

In this regard, it is estimated that in the year 1976, on projects which cost in excess of $20 million, it will not be uncommon that 50 to 100 letters will be received and an identical number sent to the prime contractor monthly!

Lastly, better listening habits can often facilitate communications.

SUMMARY

Putting it together...that's what AECM work is, and the tools presented throughout this book are ours to use. The author has attempted throughout the book to introduce functional documents, checklists, and guides that are practical and can serve as general aids in AECM work. This is intended to allow the reader to abstract these documents from the text as necessary or desirable and to adapt them to his purposes.

By attempting to capture specifics and details which can be so valuable to practitioners and newcomers alike in the AECM field, only an introduction to, or general overview of, the elusive AECM field could be developed. Obviously this AECM field is more properly the subject of numerous texts or an entire curriculum, rather than of a single text.

Nevertheless, if nothing else is recognized or appreciated as a result of this representation than the wide range of activity and the diverse character of the elements of AECM work, then this basic handbook has served a useful purpose.

A complete and well-executed project doesn't just happen...successful construction operations do not always follow naturally from design...efficient, economical, and aesthetically acceptable design doesn't come about by accident.

We, as an AECM team, must "put it all together."

Appendix

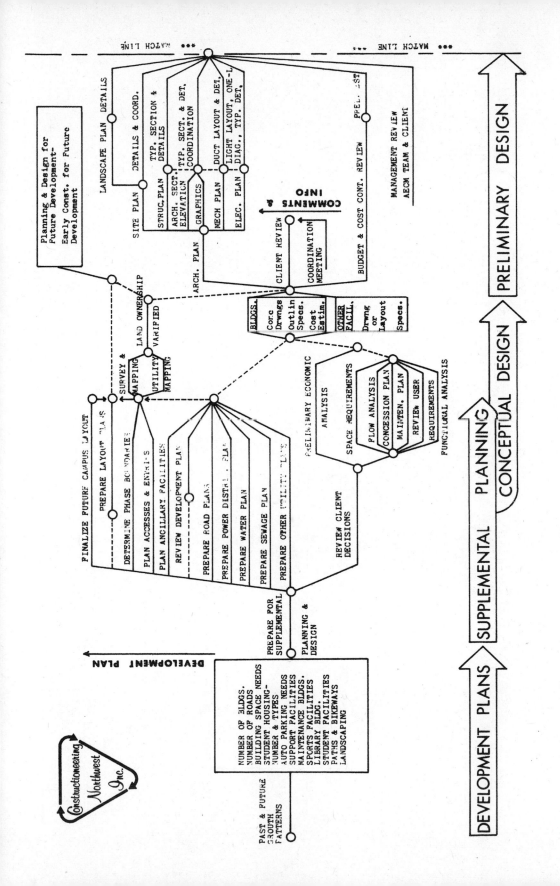

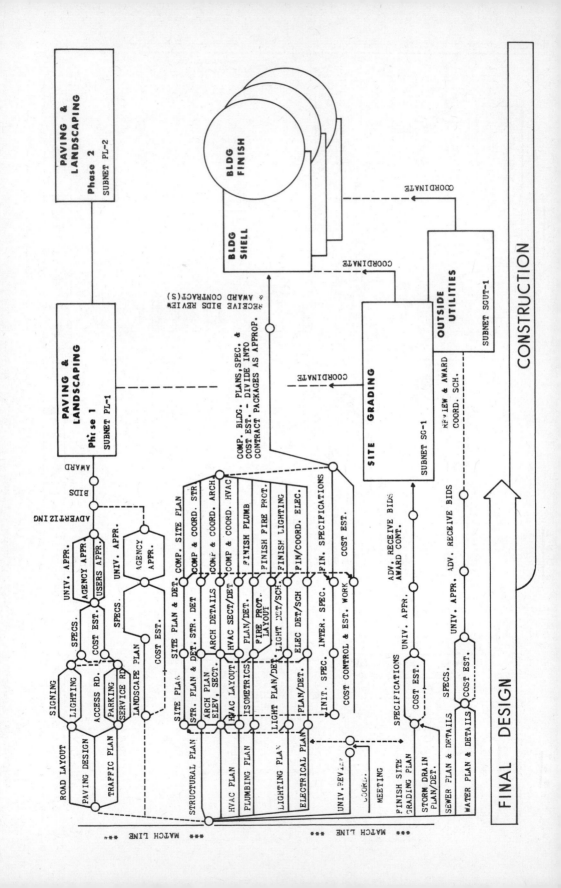

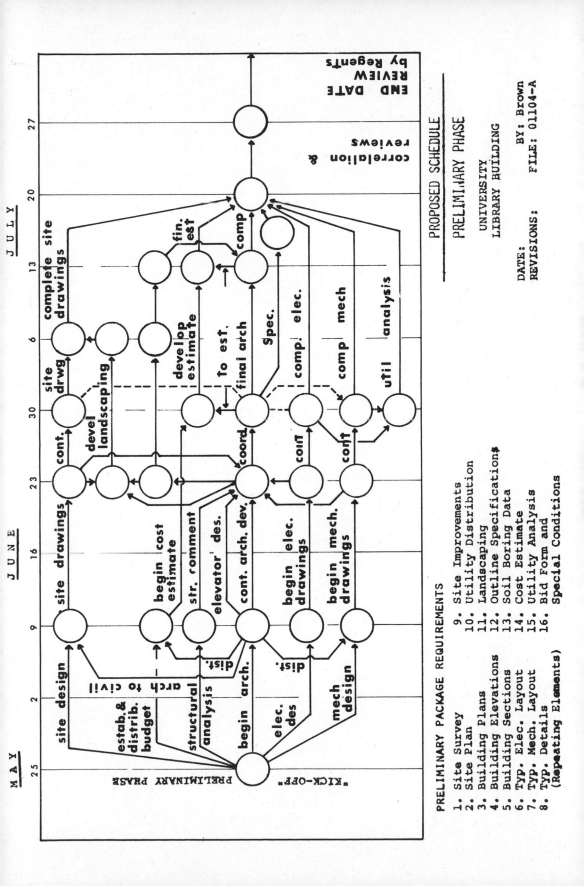

PROPOSED SCHEDULE

PRELIMINARY PHASE

UNIVERSITY
LIBRARY BUILDING

DATE: BY: Brown
REVISIONS: FILE: 01104-A

PRELIMINARY PACKAGE REQUIREMENTS

1. Site Survey
2. Site Plan
3. Building Plans
4. Building Elevations
5. Building Sections
6. Typ. Elec. Layout
7. Typ. Mech. Layout
8. Typ. Details
   (Repeating Elements)
9. Site Improvements
10. Utility Distribution
11. Landscaping
12. Outline Specifications
13. Soil Boring Data
14. Cost Estimate
15. Utility Analysis
16. Bid Form and
    Special Conditions

Index